This unique volume provides a complete reference on variable stars. It presents a wealth of typical light and colour curves (more than 200 in all) to allow identification, together with a detailed and up-to-date description of each subclass (giving the observational characteristics, historic background and current understanding of the astrophysical processes responsible for the variability).

The editors, together with seven other world experts, have created a unique pictorial atlas of variable stars. In the first chapter they give a clear introduction to the nomenclature and classification of the light curves of variable stars, and to photometric systems and photometric accuracy. In the remaining chapters they provide a detailed account of each subclass of variable star in turn from eruptive, pulsating, rotating and cataclysmic variables, through to eclipsing binary systems and X-ray binaries. Specific variable stars, types and classes of variables, together with key astrophysical terms can be quickly and easily located in the book by means of detailed object-name and subject indexes.

This comprehensive and up-to-date volume provides an essential reference for all those interested in variable stars – from researchers and graduate students through to dedicated amateurs.

LIGHT CURVES OF VARIABLE STARS

LIGHT CURVES OF VARIABLE STARS

A Pictorial Atlas

Edited by
C. STERKEN
University of Brussels (VUB), Brussels, Belgium

C. JASCHEK
Centre de Données Astronomiques de Strasbourg (CDS), Strasbourg, France

CAMBRIDGE UNIVERSITY PRESS
Cambridge, New York, Melbourne, Madrid, Cape Town, Singapore, São Paulo

Cambridge University Press
The Edinburgh Building, Cambridge CB2 2RU, UK

Published in the United States of America by Cambridge University Press, New York

www.cambridge.org
Information on this title: www.cambridge.org/9780521390163

First published 1996
This digitally printed first paperback version 2005

A catalogue record for this publication is available from the British Library

Library of Congress Cataloguing in Publication data

Light curves of variable stars: a pictorial atlas / edited by C. Sterken, C. Jaschek.
p. cm.
Includes bibliographical references and index.
ISBN 0 521 39016 8
1. Variable stars – Light curves – Atlases. I. Sterken, C. (Christiaan) II. Jaschek, Carlos.
QB836.L54 1996
523.8′442–dc20 96-1583 CIP

ISBN-13 978-0-521-39016-3 hardback
ISBN-10 0-521-39016-8 hardback

ISBN-13 978-0-521-02024-4 paperback
ISBN-10 0-521-02024-7 paperback

In grateful appreciation for
meticulous bookkeeping and dissemination
of information on variable stars
for decades, in most difficult circumstances

we dedicate this book to

the Editors

P.P. Parenago,† B.V. Kukarkin,† P.N. Kholopov,† N.N. Samus'

and Authors

N.M. Artiukhina, O.V. Durlevich, Yu.I. Efremov, Yu.N. Efremov, V.P. Federovich,
M.S. Frolov, V.P. Goranskij, N.A. Gorynya, E.A. Karitskaya, E.V. Kazarovets,
N.N. Kireeva, N.P. Kukarkina, N.E. Kurochkin, N.A. Lipunova, G.I. Medvedeva,
E.N. Pastukhova, N.B. Perova, G.A. Ponomareva, Yu.P. Pskovsky, A.S. Rastorguev,
S.Yu. Shugarov, T.M. Tsvetkova

of the

General Catalogue of Variable Stars

and to

the Editors

L. Detre,† B. Szeidl, L. Szabados, K. Oláh

of the

Information Bulletin on Variable Stars

† deceased

Contents

Contributors

Hilmar W. Duerbeck
Astronomisches Institut, Westf. Wilhelms-Universität Münster, Wilhelm-Klemm-Str. 10, D-48149 Münster, Federal Republic of Germany

Michael W. Feast
Department of Astronomy, University of Cape Town, Rondebosch 7700, South Africa

Douglas S. Hall
Dyer Observatory, Vanderbilt University, Nashville, Tennessee 37235, U.S.A.

Carlos Jaschek
Centre de Données Astronomiques de Strasbourg (CDS), Observatoire de Strasbourg, 11, rue de l'Université, 67000 Strasbourg, France

Gérard Jasniewicz
Centre de Données Astronomiques de Strasbourg (CDS), Observatoire de Strasbourg, 11, rue de l'Université, 67000 Strasbourg, France

Joachim Krautter
Landessternwarte, Heidelberg-Königstuhl, Königstuhl 12, 69117 Heidelberg, Federal Republic of Germany

Christiaan Sterken
University of Brussels (VUB), Pleinlaan 2, 1050 Brussels, Belgium

Nikolaus Vogt
Astrophysics Group, Universidad Católica, Casilla 104, Santiago 22, Chile

Patricia A. Whitelock
South African Astronomical Observatory, P.O. Box 9, Observatory 7935, Cape, South Africa

Preface

During the preparation of the observing programme of the TYCHO project on board the HIPPARCOS mission we started thinking about the large number of new variable stars that would be discovered. And since the TYCHO experiment yields only a scanty number of scattered measurements of each star during the life time of the satellite, it is immediately evident that one will encounter the problem of recognising the type or class of variability to which the variable star belongs. Such classification is – even with abundant data – not a trivial task, since many variable stars have light curves which, at first sight, look very similar. In addition, proper classification needs much more than a good-looking light curve, since luminosity and effective-temperature photometric indices also play a role, as well as miscellaneous data obtained with apparatus that are complementary to photometric instruments.

We thought to get some help by looking for standard light curves of typical variable stars that would be used as a template during the process of classification. We discovered then, with some surprise, that a compilation of typical photoelectric light curves of variable stars has never been published, nor does there exist a concise compendium of photometric properties of groups and classes of variables. What can be found, instead, is a large number of detailed morphological descriptions and numerous photometrically-incompatible photographic and visual light curves, scattered over many books and journals.

So, we decided to fill this gap and we started the compilation of typical light curves in a format that enables quick recognition of the pattern of variability. And we looked for a concise description explaining the physical processes that cause the observed variability, or an indication of which phenomena are not well understood. It was immediately obvious to us that no single person would be able to complete such a job, so the help of half a dozen experts was sought, and this book is the result of our joint efforts.

We first intended to present all data (light and colour curves) in one single photometric system, and to represent the data with a single graphical software package. However, as the light curve data were coming in, we realised that it was just impossible to combine all photometric data into one single homogeneous data set. Moreover, we learned with dismay that for many light curves published less than a decade ago, the only remnant data are the graphs themselves, the original data having been lost forever – indeed a vivid demonstration that archiving of astronomical data is still in its infancy.

We have thus chosen the most representative photoelectric light and colour curves that could be found, and we have reproduced them, together with photographic light curves, where necessary. For some stars with large-amplitude light variations on very long time scales, we have illustrated the photoelectric light curve with a visual light curve, based mostly on data that were kindly supplied by the AAVSO. All these visual-magnitude graphs are displayed on approximately the same magnitude scale along the Y-axis.[1] The time axis, however, varies from one curve to another, and gives an expressive illustration of the monumental work achieved by the joint efforts of hundreds of amateur astronomers all over the world, almost throughout the entire 20th century, and equally to a most wonderful archiving task carried out by the AAVSO.

The book is organised into seven chapters. Chapter 1 provides a general introduction. Chapters 2 to 7 deal with the subclasses and light curve properties of each of the six main groups of variables as defined in the Fourth edition of the *General Catalogue of Variable Stars* (*GCVS*). Each *GCVS* group is subdivided in several subclasses, and these subclasses are the topic of a separate Section. Each Section consists of a short description of the type of variability that is discussed, a representation of the historical background, and finally of a number of light and colour curves for several notorious stars member of the class. We deviate from the *GCVS* tradition of using only roman letters (like BCEP for the β Cep class) since that convention was made for no other reason than a practical one in the printing of the *GCVS* catalogues, and there is, today, no technical justification for preserving that convention.

We are convinced that this book will not only be useful during the analysis of TYCHO data, but also that other projects will profit from it. In particular, the growing number of Automatic Photometric Telescopes (APTs) being commissioned worldwide, the increasing number of serendipitous discoveries of variable stars in all kinds of CCD imaging projects, and the searches in data banks and archives continuously yield light curves to non-specialists. In addition to being a guide on how to read light curves, this book will also be a useful tool for the astronomy student and researcher, a textbook for astronomy

1. except for for light curves of supernovae, where the large magnitude range would yield graphs exceeding the technically available space on a single page

teachers and thesis supervisors, and also a guide for amateur astronomers. The search for specific variable stars, types and classes of variables, and astrophysical keywords, is made very easy through a most detailed name and subject index.

We also believe that this book comes at the right time. Variable-star research has been done so far using 'small' telescopes – that is, telescopes with apertures from about 150 cm down to 50 cm, and even below. Such research, for reasons of the photon budget, was mostly confined to our own galaxy. For some years, galactic work, in general, has become less popular and extra-galactic research is becoming the fashion. Variable-star work, in particular, is looked at with some disdain, and the 'small-telescope' tool is mentally associated with small science. Today it is mostly forgotten that the persistent study of variable stars has led us to answers on many questions on stellar structure and on stellar-atmosphere physics.

A very important task, carried out almost invisibly through nearly half a century, has been the bookkeeping and the classification of newly-discovered variables, along with the dissemination of information, both in catalogue form (*GCVS*) and in information bulletins (*IBVS*). Compilations that have taken man-centuries of labour in very difficult circumstances by dozens of colleagues, almost exclusively at the Sternberg Astronomical Institute in Moscow and at the Konkoly Observatory in Budapest, for no other return than the satisfaction in serving the whole astronomical community. It is for this reason that we dedicate this work to all those who have contributed and are still contributing to this task.

C. Sterken
C. Jaschek

Acknowledgements

We wish to thank a number of persons without whose help this book could not have been completed. We are indebted to the Staff of Sonneberg Observatory for many years of collaboration on the matter of bibliography of variable stars. Special thanks go to Dr J. Mattei, Director of the American Association for Variable Star Observers, who provided us with several graphs based on archived AAVSO data, and to Mr Albert Jones (RASNZ and Carter Observatory, Wellington, New Zealand) for the permission to use some of his unpublished visual estimates.

We thank Dr M. Crezé, Director of Strasbourg Observatory, and the staff of Strasbourg Observatory for their help. We emphasize the support by Messrs C. Schohn and J. Marcout, who made the photographic reproductions of a number of light curves, and also the help by Mr Jan Van Mieghem of the University of Brussels, who edited and upgraded some graphs and figures. We are grateful to Mrs M.J. Wagner and the Strasbourg Librarian Mrs M. Hamm who helped on the bibliography. Bibliography not available at Strasbourg was kindly provided by Mrs J. Vin, Librarian of the Haute Provence Observatory. We very much appreciate the helpful support by Mrs M. Vargha, Librarian of Konkoly Observatory in Budapest, and by Mr P. Dale, Librarian of the Royal Observatory in Brussels (Uccle).

C.S. expresses his gratitude to the Belgian Fund for Scientific Research for supporting several research projects directly related with his contributions to this book.

Journal abbreviations

Acta Astron.	*Acta Astronomica*
Ann. Tokyo Obs.	*Annals of the Tokyo Observatory*
Ann. Rev. Astr. Ap.	*Annual Review of Astronomy and Astrophysics*
Ap. J.	*Astrophysical Journal*
Ap. J. Lett.	*Astrophysical Journal Letters*
Ap. J. Suppl.	*Astrophysical Journal Supplement Series*
Ap. Sp. Sci.	*Astrophysics and Space Science*
Ark. f. astron.	*Arkiv för Astronomi*
Astr. Ap.	*Astronomy and Astrophysics*
Astr. Ap. Rev.	*Astronomy and Astrophysics Reviews*
Astr. Ap. Suppl.	*Astronomy and Astrophysics Supplement Series*
Astron. J.	*Astronomical Journal*
Astron. Nachr.	*Astronomische Nachrichten*
Astron. Soc. Pac. Conf. Ser.	*Astronomical Society of the Pacific Conference Series*
Bull. AAS	*Bulletin of the American Astronomical Society*
Bull. Astron. Inst. Czech.	*Bulletin of the Astronomical Institutes of Czechoslovakia*
Bull. Astron. Inst. Neth.	*Bulletin of the Astronomical Institutes of the Netherlands*
Bull. Astron. Inst. Neth. Suppl.	*Bulletin of the Astronomical Institutes of the Netherlands Supplement Series*
Bull. Crim. Astr. Obs.	*Bulletin of the Crimean Astrophysical Observatory*
Bull. Inf. CDS	*Bulletin d'Information Centre de Données Stellaires*
Cape Ann.	*Annals of the Cape Observatory*
GCVS	*General Catalogue of Variable Stars, Fourth Edition*
Fund. Cosmic Phys.	*Fundamental Cosmic Physics*
Harvard Bull.	*Harvard Bulletin*
Hvar Obs. Bull.	*Hvar Observatory Bulletin*
IAU Circ.	*International Astronomical Union, Circular*

IAPPP Comm.	*International Amateur Professional Photoelectric Photometry Communications*
Inf. Bull. Var. Stars (IBVS)	*Information Bulletin on Variable Stars*
JAAVSO	*Journal of the American Association of Variable Star Observers*
JAD	*Journal of Astronomical Data*
JRAS Canada	*Journal of the Royal Astronomical Society of Canada*
Lect. Notes Phys.	*Lecture Notes in Physics*
Lick Obs. Bull.	*Lick Observatory Bulletin*
Mem. R. Astron. Soc.	*Memoirs of the Royal Astronomical Society*
Mem. Soc. Astron. Ital.	*Memorie della Societa Astronomica Italiana*
Messenger	*The Messenger*
Mitt. Ver. Sterne Sonneberg	*Mitteilungen über Veränderliche Sterne, Berlin-Babelsberg & Sonneberg*
MNASSA	*Monthly Notes of the Astronomical Society of Southern Africa*
MNRAS	*Monthly Notices of the Royal Astronomical Society*
Observatory	*The Observatory*
Perem. Zvezdy	*Peremennye Zvezdy*
Proc. Astron. Soc. Aust.	*Proceedings of the Astronomical Society of Australia*
Proc. Amer. Acad. Arts Sci.	*Proceedings of the American Academy of Arts and Sciences*
Publ. Astron. Soc. Japan	*Publications of the Astronomical Society of Japan*
Publ. Astron. Soc. Pac.	*Publications of the Astronomical Society of the Pacific*
SAAO Circ.	*South African Astronomical Observatory Circular*
Sky Telesc.	*Sky and Telescope*
Space Sci. Rev.	*Space Science Reviews*
Trans. IAU	*International Astronomical Union Transactions*
Zeitschr. f. Astrophys.	*Zeitschrift für Astrophysik*

Acronyms and abbreviations

AG	Astronomische Gesellschaft
AAVSO	American Association of Variable Stars Observers
AFOEV	Association Française des Observateurs d'Etoiles Variables
AGB	asymptotic giant branch
APT	Automatic Photometric Telescope
BAA	British Astronomical Association
BCSVS	*Bibliographic Catalogue of Suspected Variable Stars*, Roessiger & Braeuer (1993, 1994)
BCVS	*Bibliographic Catalogue of Variable Stars*, Roessiger & Braeuer (1994)
BD	*Bonner Durchmusterung*
CCD	charge coupled device
CDS	Centre de Données Stellaires, Observatoire de Strasbourg
CP	chemically peculiar
CV	cataclysmic variable
EROS	Expérience de Recherche d'Objets Sombres
ESO	European Southern Observatory
ESA	European Space Agency
ftp	file transfer protocol
GCVS	*General Catalogue of Variable Stars* 4th ed., Kholopov *et al.* (1985a)
He 3	Henize catalog number, *see* Section 1.2
HIPPARCOS	HIgh Precision PARallax COllecting Satellite
HMXRB	high-mass X-ray binaries
H–R	Hertszprung–Russell (diagram)
IAPPP	International Amateur Professional Photoelectric Photometry
IAU	International Astronomical Union
IBVS	*Information Bulletin on Variable Stars*
IP	intermediate polars

IUE	International Ultraviolet Explorer
LMXRB	low-mass X-ray binaries
LBV	Luminous Blue Variable
LPV	Long-Period Variable (Mira variable)
LMC	Large Magellanic Cloud
LTPV	Long-Term Photometry of Variables project
LTSV	Long-Term Spectroscopy of Variables project
MACHO	Massive Compact Halo Object
MLA	maximum light amplitude
NSV	*New Catalogue of Suspected Variable Stars*, Kukarkin *et al.* (1982)
OGLE	Optical Gravitational Lensing Experiment
OPAL	opacity code developed at Lawrence Livermore National Laboratory
PL	period–luminosity relation
PLC	period–luminosity–colour relation
PMS	pre-main sequence
PNNV	variable planetary nebula nuclei
R	Ratcliffe catalog number, *see* Section 1.2
RASNZ	Royal Astronomical Society of New Zealand
ROSAT	Röntgen Satellite
S	Henize catalog number, *see* Section 1.2
SAT	Strömgren Automatic Telescope
SIMBAD	Set of Identifications, Measurements and Bibliography of Astronomical Data
SMC	Small Magellanic Cloud
SPB	slowly-pulsating B stars
TYCHO	photometric instrument on board of HIPPARCOS
WD	white dwarf
WR	WR star, also WR catalog number, *see* Section 1.2

1
Introduction

1.1 Variable stars

G. Jasniewicz

All stars display variations of brightness and colour in the course of their passage through subsequent stages of stellar evolution. As a rule, however, a star is called *variable* when its brightness or colour variations are detectible on time scales of the order of the mean life time of man. The variations may be periodic, semi-periodic or irregular, with time scales ranging from a couple of minutes to over a century. It is this kind of variable star which is the topic of this book. The typical time scale, the amplitude of the brightness variations, and the shape of the light curve can be deduced from photometric observation, and those quantities place the star in the appropriate class. For example, a star of the UV Ceti type typically has brightness variations (the so-called flares) of several magnitudes in an interval of time as short as a few minutes, whereas a Cepheid shows periodic variations of about one magnitude in a time span of several days. However, spectral type, luminosity class and chemical composition are complementary important spectroscopic parameters that are needed for classifying variable stars according to the *origin* of their variations.

Due to stellar evolution, a variable star discloses long time-base variations of its amplitude and of its typical duration of the cycle of variability: for a short-period binary star, variations of the period and of the times of minimum or maximum of the light curve can be explained by modifications in the mass transfer between both components; for a δ Scuti type star, a change of the period can be due to a change of the radius by the evolution of the star from the dwarf stage to the giant stage. On the other hand, for the RV Tauri type stars, the origin of the observed abrupt changes of period is still not understood.

Thus, variations on long time scales of the typical photometric parameters of variable stars give us information on the physical processes which are responsible for the observed brightness and colour variations. Therefore it is of crucial importance that astronomers undertake long-term observing programs of variable stars and also archive all photometric measurements done in the past and today.

Archiving of variable-star photometric data

Proper archiving is crucial for a complete understanding of the mechanisms of variability. Archives can also help to solve ambiguities: it can happen that period analysis on a set of data yields a period different than a previously published one, simply because the methods and the understanding of statistics of period analysis have evolved. Unfortunately, access to original data today is not a simple issue in our international astronomical community. Indeed, if a list of data is very extensive, the majority of the astronomical journals will not accept it for publication, not even in their Supplement Series. Therefore, observers themselves do archive their own data and, occasionally, make them available for public use. Thus, in order to prepare the light curves for this book, we have asked authors for their unpublished data. Not entirely to our surprise, no more than one out of every five responded positively.

We emphasize here that, in order to avoid the definitive loss of unpublished observations of variable stars, the International Astronomical Union (IAU) has asked its Commissions 27 (Variable Stars) and 42 (Close Binary Systems) to archive the photometric observations of variable stars. These Archives replace lengthy tables in scientific publications by a single reference to the archival file number, and they also contain many valuable observations which have never been used for scientific publication. M. Breger, at Vienna Observatory, was for many years the coordinator of the IAU Archives; today the conservator is E. Schmidt of the University of Nebraska at Lincoln. Electronic storage and retrieval of photometric data from the Archives can be done by communicating with the Centre de Données de Stellaires (CDS); some files can be electronically obtained free of charge from CDS. Copies of all existing files can be obtained in paper form from three archives, viz. from P. Dubois (CDS, France), from P.D. Hingley (Royal Astronomical Society, Great Britain) and from Y.S. Romanov (Odessa Astronomical Observatory). Lists of the available files in the Archives are regularly announced in the *Information Bulletin on Variable Stars* (*IBVS*, Konkoly Observatory, Budapest) and the *Bulletin d'Information du CDS* (see Jaschek & Breger, 1988). Detailed reports on the file contents are published by Breger (1988).

We also mention that several amateur associations, such as the Association Française des Observateurs d'Etoiles Variables (AFOEV), the American Association of Variable Stars Observers (AAVSO), the Royal Astronomical Society of New Zealand (RASNZ) and the British Astronomical Association (BAA) have their own archives in computer-readable form – the bulk of these data, however, are visual estimates by eye.

Variable-star monitoring programs

A large variety of observing programs of variable stars is presently in progress all over the world. An important part of all major astronomical journals is devoted to individual stars which display regular or erratic variations of brightness. Announcements of newly discovered novae and supernovae or of drastic changes in cataclysmic stars are published by the Central Bureau for Astronomical Telegrams (Brian G. Marsden) that depends on IAU Commission 6. Observations of variable stars can be quickly published by *IBVS* and, since 1995, also in purely electronic form in *The Journal of Astronomical Data* (*JAD*, TWIN Press, Sliedrecht, The Netherlands).

The importance of studies of stellar variability on long time scales has been emphasized above. Long-term observing programs of variable stars are in progress in major observatories. One such program, Long-Term Photometry of Variables (LTPV), has been operating since 1982 at the European Southern Observatory (ESO) in Chile (Sterken, 1983), with archived data at SIMBAD. This book contains several graphs made with these data ('based on LTPV data', see, for example, Fig. 2.1), the data tables can be found in Manfroid *et al.* 1991a, 1991b, 1994a, 1994b, and in Sterken *et al.* 1993a, 1993b, 1995b, 1995c. More than a few advantages and problems of coordinated campaigns and of long-term monitoring have been elucidated by Sterken (1988, 1994). The benefits of ground-based support of space observations of variable stars are obvious: in the case of very hot stars (white dwarfs, central stars of planetary nebulae, etc.), the bulk of radiated energy is at the short wavelengths and is not detectable from earth. For example, simultaneous IUE and ground-based photometric observations of the binary central star of the planetary nebula Abell 35 have strongly contributed to clarify the nature of the so-called 'Abell 35'-type objects: such binaries experience chromospheric activity as RS CVn-like binaries (Jasniewicz *et al.* 1994b); besides, the late-type giant stars in the nuclei display characteristics common to the FK Comae stars (Jasniewicz *et al.* 1987, 1994a).

The advantages of observations collected at discrete geographical longitudes are also clear because quasi-contiguous sequences of measurements can in principle be collected, and then the amplitudes of the alias periods resulting from the period searches can be significantly reduced.

The contributions by amateur variable-star observers also play an important role in the continous monitoring of variable stars; that is especially the case for cataclysmic stars and long-period variables such as the Mira Ceti type stars. In this book the behaviour of variable stars on very-long time scales is illustrated by means of light curves kindly provided by the AAVSO and by individual amateurs.

Searching for new variable stars

Several ground-based and space projects will, in the coming years, considerably increase the number of known variables. We focus on two examples, viz. the HIPPARCOS/TYCHO projects, and the search for massive compact galactic objects.

HIPPARCOS (HIgh Precision PARallax COllecting Satellite) was launched by the European Space Agency (ESA) on August 8, 1989. The HIPPARCOS mission is now accomplished, and processing of the collected astrometric and photometric data (about 100 000 measurements) is progressing.

An ambitious photometric program is the TYCHO photometric survey with the satellite HIPPARCOS. The TYCHO program involves photometric observation in two colours of about 1 000 000 stars across the sky. Photometry is based on the star-mapper signal and two detectors at effective wavelengths $\lambda 428$ and $\lambda 534$ nm. The two filters B_T and V_T of the TYCHO experiment differ slightly from the standard Johnson B and V filters (Mignard *et al.* 1989, see Fig. 1.1). Several thousand confirmed standard stars were used for calibrating the photometric system. Every program star was observed about 100 times during the satellite's life time (January 1990 to March 1993); thus the TYCHO project will yield a substantial number of new variable stars. However, for stars fainter than $V_T = 10$, the survey will become non-homogeneous. In using empirical detection probabilities and statistics of known variable stars, various authors (Jaschek 1982, Mauder & Høg 1987, Halbwachs 1988) have estimated the numbers of new variable stars expected to be found in the TYCHO photometric survey. According to Mauder & Høg (1987) the number of variables brighter than $B = 11$ would be at least doubled for most types of variability (see Table 1.1). Several hundred Cepheids and thousands of eclipsing binaries should be discovered by means of TYCHO. A methodological strategy for analyzing the photometric data of the TYCHO space experiment has been announced by Heck *et al.* (1988). In order to identify new variable stars, various statistical algorithms will have to be applied, especially algorithms

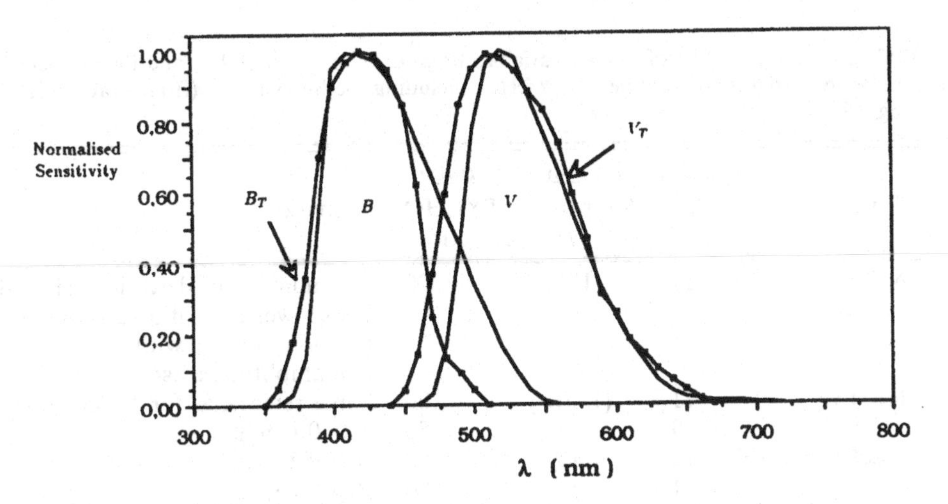

Figure 1.1 Normalised response curves for TYCHO B_T, V_T compared to $B; V$ (from Mignard *et al.* 1989).

which allow to distinguish a variable signal from noise, and also algorithms for period analysis. The next step will be to assess the type of variability of each newly-discovered variable star. The present book can be a helpful tool for this task because it brings an exhaustive and up-to-date compilation of light curves of all known types of variable stars.

A second key project is the ground-based microlensing experiments dedicated to the search for massive compact objects (black holes, brown dwarfs, etc.) in our galactic halo. According to Paczýnski (1986) massive dark bodies might act as gravitational microlenses, temporarily amplifying the apparent brightness of background stars in nearby galaxies. The most widely known microlensing experiments are the Polish-American OGLE (Optical Gravitational Lensing Experiment) project, the American-Australian MACHO (Massive Compact Halo Object) project and the French EROS (Expérience de Recherche d'Objets Sombres) project. Besides the discovery of microlensing events, a very interesting sub-product of all these experiments is the discovery of numerous variable stars among stars in the Large Magellanic Cloud (MACHO and EROS) and towards the Galactic Bulge (OGLE). An EROS catalogue of new eclipsing binary stars in the bar of the Large Magellanic Cloud has already been published (Grison *et al.*, 1994) and many other papers are in preparation. A lot of insights into stellar evolution should follow from these newly-discovered variables, and it will be a very rewarding enterprise to store the entire resulting data set in various locations for archival purposes.

Table 1.1 Number (N) of known variables brighter than the limit B_{lim} and the predicted number of variables detectable by TYCHO, including the known ones (from Mauder & Høg, 1987)

Type	B_{lim}	Number of variables: Known N	TYCHO detections	Remarks
δ Cep	11	173	1000	The number of detections should be a lower limit if a substantial fraction of stars with low amplitude exists
RR Lyr	11	(155)	100	Same remark as for δ Cep stars
δ Sct	9	8	> 50	$A \geq 0.1$ mag
δ Sct	10	13	> 50	$A \geq 0.2$ mag
α^2 CVn	11	14	> 100	$A \geq 0.1$ mag
EA	11	710	2000	About 80% of all EA detectable
EB	11	706	1500	Probably complete detection by TYCHO
EW	11	88	100	If the expected number of EW stars with low amplitude is correct, there could even exist about 1000 such objects
LPV, SR, IRR	11	2400	some 100	Some 1000 could perhaps be found

1.2 Nomenclature of variable stars

C. Sterken

For a working definition of the concept of variable star, we refer to Section 1.1; for a slightly too broad definition, we mention 'A star is called variable, when it does not always appear to us with constant brightness...' (Schiller 1923). If one of the brighter objects of a constellation turned out to be variable, its Bayer notation (Greek or Roman letter followed by the genitive of the Latin constellation name) was employed as the name of the variable. The visual observations leading to the compilation of the *Bonner Durchmusterung (BD)* by Argelander in the middle of the 19th century led to the discovery of a substantial number of new variable stars. Thus, Argelander introduced a specific nomenclature for those variables that had no Bayer number in a constellation by designating these stars by the Roman capital letters R, S, T, U, V, W, X, Y and Z attached to the Latin name of the constellation. With the introduction of the photographic plate, the number of variable stars in each constellation quickly exceeded the capacities of Argelander's scheme, and the Astronomische Gesellschaft (AG) extended the scheme by adopting double

Roman letters RR to RZ, SS to SZ, etc. That extension brought the number of possible variable star names per constellation from 9 to 54. When that number turned out to be insufficient, AA to AZ, BB to BZ, etc. (but omitting J) was used. That second extension brought the number of stars to be assigned a name to 334 per constellation. According to Hoffmeister (1984), the Dutch astronomer Nijland proposed that a new and uniform system of nomenclature be introduced, and suggested that R, S, ... QZ be replaced by V 1, V 2, ... V 334 with subsequently discovered variables being labeled V 335, V 336, etc., the highest number then referring to the exact number of variables known in that specific constellation.[1] That suggestion was not entirely accepted for reasons of convenience, since many names of variables, like U Gem and RR Lyr, had already been assigned a specific class of variables, but it became in widespread use for labeling all variables that cannot be named in the extended Argelander nomenclature, i.e. from V 335 onwards. In principle, this system can be used for any possible number of variables that are discovered in a single constellation. However, with the advent of the discovery of larger and larger numbers of variables by space- and ground-based surveys (see Section 1.1), a discussion has been opened at IAU Commission 25 on the question whether such a scheme should be maintained, or whether one should produce a new scheme based on a system that directly and unequivocally refers to the position of the star in a reference coordinate system.

For novae a special nomenclature was used for a long time: older novae are designated by the label *Nova* followed by the constellation and the year in which they appeared (to the observer), for example Nova Cygni 1600. When necessary, the running number of the nova that erupted in this constellation that year is added. This system of labeling is still used as a *provisional* designation until the nova has assigned a definite variable-star designation (thus, for example, Nova Vul 1968, number 2 has become LU Vul). Since the assignment of variable-star designations to novae was not carried out in the first decades of the 20th century, the chronology and the variable-star designations do not run parallel, for example, AT Sgr = N Sgr 1900, V 737 Sgr = N Sgr 1933, V 1016 Sgr = N Sgr 1899, V 1148 Sgr = N Sgr 1948.

A similar nomenclature is used for supernovae, where the year of appearance and a sequel character is used, for example, Supernova 1987a. However, supernovae – as extragalactic objects – never receive another variable-star designation (with the exception, of course, of well-known galactic supernovae, such as SN 1572 Cas = B Cas, SN 1604 Oph = V 843 Oph).

Alternative designations for specific stars used in this book are, besides the

1. See also André (1899).

HD and HR numbers: He 3 (Henize 1976), R ('Ratcliffe Catalog': R 1–R 50 in the SMC, R 51–R 158 in the LMC, Feast *et al.* 1960), S (Henize 1956), and WR (see also van der Hucht *et al.* 1981). As in the case of supernovae, the main reason for such alternative nomenclature is that these stars are extragalactic variables, to which no variable star designation is assigned.

The specific problem of assigning a unique name to a variable star is, of course, only part of the more universal problem of labeling celestial objects in a clear and internally consistent system. Variable-star observers using two-dimensional detectors quite often discover numerous variable stars to which neither a variable-star name, nor another designation had been given previously. This, especially, is the case when observing open clusters using such detectors with high quantum efficiency. In addition, such investigations necessitate the earmarking of quite a number of non-variable comparison and standard stars. Instead of extending existing schemes, some authors, unfortunately, find it necessary to replace previously assigned designations (even HD and HDE numbers allocated more than half a century ago) by private numbering schemes and acronyms that involve the initials of their own names.

In order to avoid the practice of creating new appellations out of thin air spreading like wildfire, the IAU has issued a public notice '*Specifications concerning designations for astronomical radiation sources outside the solar system*'[1] that shows how to refer to a source or how to designate a new one.

For general information, in particular about existing designations, one should consult Fernandez *et al.* (1983), Lortet & Spite (1986), Dickel *et al.* (1987), Jaschek (1989), and Lortet *et al.* (1994).

1.3 The classification of variable stars

C. Sterken

Variable-star specialists know that it is a very difficult enterprise to construct a physically sound and consistent taxonomy of classes and types of variable stars. One of the very first attempts at classification of variable stars was the scheme of Pickering (1881), who recognised five classes, viz. *new stars [novae], long-periodic variables, irregular variables, short-period variables* and *eclipsing variables.* Classification schemes based on a physical mechanism (e.g. Newcomb 1901) were doomed to fail as long as it was not understood that stellar pulsation can be the cause of variations (Plummer 1913, for reviews on pulsations and oscillations in stars we refer to Osaki & Shibahashi 1986, Osaki 1987, and to Watson 1988 for the impact of non-radial pulsations on

1. Accessible on *ftp anonymous* at node **cdsarc.u-strasbg.fr**

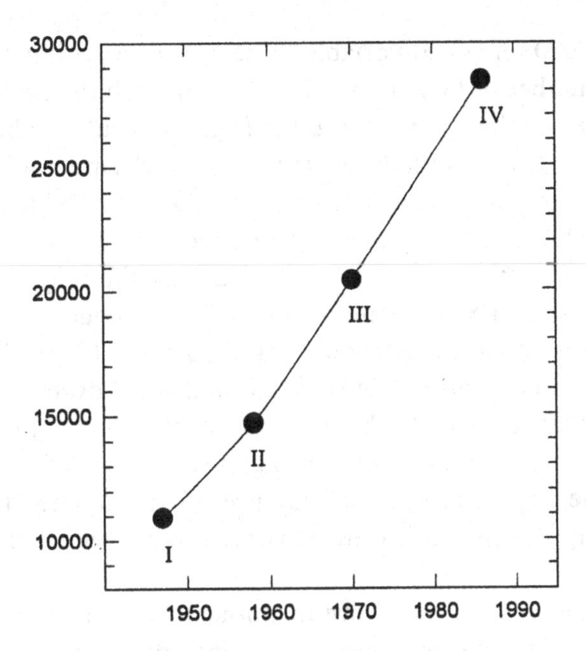

Figure 1.2 Number of variable stars listed in the four editions of *GCVS* (based on information from Samus' 1990).

flux changes). From then on, three *phyla* were always recognised: irregular variables, pulsating variables, and binary systems. Meanwhile, the classification of variable stars has undergone a change that, as Cecilia Payne-Gaposchkin puts it 'recalls the suppression of the Linnaean system by the modern system of botanical classification' (Payne-Gaposchkin 1978).

A first classification outline was published in 1960 (*Trans. IAU* 10, 398) and was based on the recommendations of the *IAU* made in Moscow in 1958. Since then, this classification has frequently been reworked and extended. A principal source of information on the latest classification of variable stars is the 4th edition of the *General Catalogue of Variable Stars* (*GCVS*, Kholopov *et al.* 1985a, 1985b, 1987a, 1990 – Volumes I, II, III and IV).[1] *GCVS* contains a classification and miscellaneous data on almost 30 000 variable stars which had been assigned variable-star names up to 1982. Not surprisingly, *GCVS* is somehow considered as the bible of systematic placement in variable star knowledge. Supplements to the *GCVS*, the well-known *Name List of Variable Stars*, appear regularly in the Information Bulletin on Variable Stars (*IBVS*, IAU Commission 27 & 42), e.g. Kholopov et al. (1987b, 1989). We also refer to *The New Catalogue of Suspected Variables* (*NSV*, Kukarkin *et al.* 1982), which

1. Compilation and publication of *GCVS* started in 1948 under the auspices of IAU, with an intended frequency of publication of a new edition every 10 years.

contains data on almost 15 000 unnamed variable objects. Figure 1.2 illustrates the steady increase in the number of known variable stars through the past half century, as measured by the number of entries in the four consecutive editions of *GCVS*. Users of *GCVS* and *NSV* should know that these works are not just a mere compilation, but are based on a critical evaluation of all underlying data, as expressed by Samus' (1990) in the case of an announced nova (V 600 Aql – 'Nova' Aql 1946), which was proven to be an asteroid during the preparation of *GCVS*. The 5th volume of *GCVS* is the first systematic catalogue of extragalactic variables, listing data on variables in nearby galaxies, viz. LMC, SMC, Andromeda Nebula (M31), Triangulum Nebula (M33) and some dwarf galaxies. The majority of extragalactic variables belong to Cepheid types – Fig. 1.3 illustrates the type distribution in the four mentioned galaxies – except for M31, where the number of novae approaches the abundance of the Cepheid types (Lipunova 1990). For a comprehensive overview on variable stars, see Duerbeck & Seitter (1982, 1995).

Astronomers who are familiar with the above-mentioned works must notice that every newly-published catalogue or review paper introduces new classes or subclasses of variable stars, and that the classification in itself, at the same time, not only refines, but also becomes, occasionally, less and less consistent, including even the classification of the same object in different classes,[1] or the transfer of an object from one class to another (and back) as time passes. Such changes, of course, reflect the progress and understanding in the field – a progress that is not only due to more and more sophisticated detection techniques, but also to the increasing time baseline along which data are being collected. An illustrative example of the latter point is the case of the 'ex-constant star' – that is, a star that turns out to be a variable only after many years of measurements become available for analysis. Another aspect of published classification schemes is the pronounced vagueness with which some classes are being defined; the *GCVS*, for example, lists several subclasses where members are characterised by the label 'poorly studied', a fact that eventually will lead to more refinement and to the definition of new classes and subclasses. Looking over the past decade, we have seen new classes like the rapidly-oscillating Ap stars and the slowly-pulsating B stars, to name only two.

The actual classes do not have clear-cut borders, since these borders directly depend on the parameter that is being used for the classification (see also M. Feast's remarks on the classification of Cepheids in Section 3.7). If that parameter is directly related to the wavelength range in which the data are

1. Not by error, but by the fact that some variable stars are polymorphic – see, for example the Be stars (Section 3.3), where signatures of rotation, pulsation, orbital motion, eruptions and even X-rays are seen.

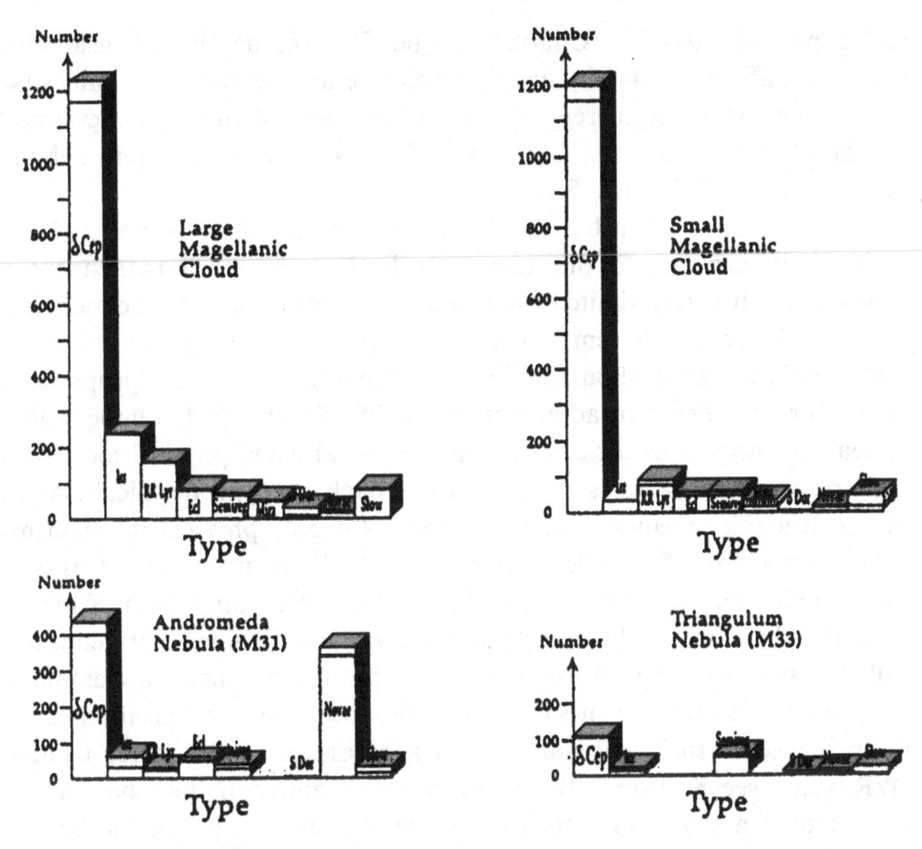

Figure 1.3 Number of variable stars of different types in four external galaxies (based on Vol. 5 of *GCVS* – Lipunova 1990).

taken, clear borders can be assigned, but if the parameter is a physical quantity like, for example, mass, a certain fuzziness must come up – not only because mass is *derived* from other observables (with their own uncertainties in turn leading to a generous error range), but also because the mass range forms an uninterrupted scale along which quantisation is being imposed.

Variable stars are, traditionally, classified in two main families: *intrinsic* variable stars, and *extrinsic* variables, respectively for stars that vary due to physical processes in the the star itself, and for stars that vary due to processes external to the star (such as rotation combined with a physical process in the star that, for example, produces spots and orbital motion – the latter case would not lead to observed variability if the observer were not at a vantage point that allows, for example, eclipses of a companion to be seen). Extrinsic variables are eclipsing variables (such as Algol-type, β Lyrae type and W UMa type, see Chapter 6), or rotational variables and pulsars (Chapter 4). Intrinsic variables are the pulsating variables (Chapter 3), and the eruptive and explosive variables (Chapters 2 and 5). Among the intrinsic variables, the most difficult to classify are the explosive and symbiotic variables (studies over limited wavelength regions have often resulted in symbiotic stars being misclassified as something else, most often as peculiar planetary nebulae, see P.A. Whitelock in Section 5.5).

Classification is often based on the most characteristic features of the light and colour curves, and this inevitably leads to a non-homogeneous result, sometimes with very limited value for a real physical classification (see, for example, J. Krautter's remarks in the introduction to Section 2.3 that states that any classification should be based on the intrinsic physical properties of the stars – for an in-depth pleading, see Hall 1995). Smith (1994), on the other hand, appeals to the photometric community to avoid using physical mechanisms in defining variable-star classes, and to fall back on a prototypical star's name in constructing schemes of classification. Not only physical mechanisms, but also the position of a variable star in the H–R diagram is used as an index for classification, especially when the position of a star falls in a region that is occupied by one of the many pulsating-variable families that delineate a so-called 'instability strip'. Another element that is often used in classification is the presence of specific emission lines in the spectrum: whereas for some classes the presence of such emission lines is a necessary condition for membership (WR stars, see Section 2.2), the presence of emission lines has also led to exclusion of a star from another class (a typical example of the latter is the case of the β Cep stars – discussed in Section 3.2 – where even the prototype occasionally has shown emission features, see Le Contel *et al.* 1981, Henrichs *et al.* 1993). Finally, there is an enormous amount of overlap due to the diversity of physical phenomena: there are pulsating stars that show eruptions, there exist spotted stars that are members of binaries, there are pre-main sequence stars that show δ Scuti pulsational variability (Kurtz & Marang 1995), and so on. And, different families of classes apparently describe the same, or at least, similar variables: for example, the family of the 53 Per/mid-B/SPB stars (Section 3.4), and the α Cyg/hypergiant and LBV/S Dor/Hubble-Sandage variables (Section 2.1). Some of the class names are even misleading (*'Luminous Blue Variables need not be blue'*, de Jager & van Genderen 1989), or most directly reflect a description based on different detection methods (53 Per and SPB stars), even very different ranges in the time base of observation (α Cyg variables and LBVs). It should be clear for any one attempting to label a variable that the distinction between one class and another is not always very obvious and that taxonomy of variable stars is not a trivial pursuit.

Few attempts towards a revolutionary new design of variable-star classification have been made. For the apprentice taxonomist, the reading of Douglas

Hall's recently proposed new classification scheme is compulsory: Hall (1995) proposes six categories corresponding to six physical mechanisms, viz. (I) eclipsing or extrinsic, (II) pulsation, (III) rotation, (IV) long-term magnetic cycles, (V) supergranulation, and (VI) transient phenomena. He illustrates this scheme with several examples, some of which are discussed in this book.

In this work we do not make any attempt to change or to improve the existing classification, instead we follow the gross classification of variables in six families as defined in the *GCVS* and in its *Supplements* as a kind of ground plan. We have confined ourselves to describing the types of variables along the main branches, overlooking the many twigs and sprigs – which often appear on different limbs in parallel – that account for the more than 130 different star types to be found in the *GCVS*. The reader must also be aware of the fact that the presented scheme does differ from the taxonomy used in the literature as, for example, presented in the *Hierarchical Display* in the Astronomy Thesaurus prepared by Shobbrook & Shobbrook (1993).

In this Introduction we first give an overview of the different types of variables which, according to *GCVS*, belong to each of these six families. For each such group we start with the definition taken *verbatim* from *GCVS*, followed by an overview table that is structured in exactly the same way as the description in *GCVS* is conceived. In addition, each table is followed by an indication about where we do not follow the *GCVS* classification (for example, by introducing a class of variables not described in *GCVS*). For more detailed information on each type of variable, we refer the reader to the corresponding sections of this book.

In the subsequent chapters and sections of this book we consistently indicate a class of variable stars either by the abbreviated name or by the full name, for example, either W UMa, or W Ursae Majoris stars. For the variables themselves, we use the abbreviated Bayer notation (for example, β Lyr) or the notation in the Argelander scheme (W UMa).

The remainder of this Section reproduces the classification schemes of the *GCVS* (their six natural groups) along with some concise information on how this book either follows the *GCVS* scheme, or deviates from it.

I. Eruptive variables

GCVS: **'We call eruptive variables the stars varying their brightness because of violent processes and flares taking place in their chromospheres and coronae.'**

In this book, the irregular (I) variables are not described as a whole. The FU Ori stars are discussed with the class of pre-main sequence (PMS) stars, viz., FU Ori, EX Ori and Herbig Ae/Be stars. As J. Krautter points out in Sec-

Table 1.2 Eruptive variables

Variables	*GCVS*				
FU Orionis	FU	T Tauri like stars			
γ Cassiopeia & Be	GCAS	Be stars			
Irregular	I	early type spectra	IA		
		mid/late type	IB		
		Orion variables	IN	INA	early type spectra
				INB	mid/late type
				INT	T Tau stars
				IN(YY)	matter accreting
		Rapid irregular	IS	ISA	early type spectra
				ISB	mid/late type
R Coronae Borealis	RCB	eruptions plus pulsation			
RS Canum Venaticorum	RS	close binaries with H and K Ca II in emission			
S Doradus	SDOR	very luminous stars (hypergiants)			
UV Ceti (flare stars)	UV	UV	KV–MV flaring on time scales of minutes		
		UVN	flaring Orion stars of UV type		
Wolf–Rayet	WR	broad emission features			

tion 2.3, the *GCVS* classification (Table 1.2), which is based on morphological photometric properties only, is not homogenous and is of rather limited value for a real physical classification. For instance, the RW Aur type variables (IS) are physically nothing other than low-mass PMS stars which are usually called T Tauri stars (INT in *GCVS*), or the FU Ori type variables (FU in *GCVS*) are also T Tauri stars which are in a special stage of their evolution (see Section 2.3 for a detailed description).

D.S. Hall, in Section 6.3, calls attention to the fact that the classification RS (RS CVn variables) appears twice in the *GCVS*: first, it appears as one type in Table 1.2, and it also appears as a type of close-binary eclipsing systems in Table 1.6. The first attachment may be misleading because the mechanism for the variability is actually rotational modulation, with the surface brightness non-uniform as a result of cool spots distributed unevenly in longitude, yet it does *not* appear as one of the types of rotating variable stars (Table 1.4). The second labeling also may be misleading because more than half of the variables classified in the *GCVS* as RS are not eclipsing.

GCVS mentions the S Dor variables, and we discuss these stars in Section 2.1. Before their first observed eruption, such stars are normally classified as α Cyg variables (ACYG in *GCVS*), therefore, this section should be read with parallel consultation of Section 3.1.

In addition, we describe the γ Cas stars as Be stars (Section 3.3).

II. Pulsating variables

GCVS: **'One calls pulsating variables the stars showing periodic expansion and contraction of their surface layers. Pulsations may be radial or non-radial.'**

Table 1.3 Pulsating variables

Variables	*GCVS*		
α Cygni	ACYG		Be–Ae pulsating supergiants
β Cephei	BCEP	BCEP	classical β Cep stars
		BCEPS	short period β Cep stars
Cepheids	CEP	CEP	radially pulsating F Ib–II stars
		CEP(B)	double mode pulsators
W Virginis	CW	CWA	population II, Period > 8^{d}
		CWB	population II, Period < 8^{d}
Classical Cepheids	DCEP	DCEP	classical Cepheids (pop. I)
		CEP(S)	classical Cepheids (overtone)
δ Scuti	DSCT	DSCT	A0–F5III V pulsating stars
		DSCTC	low amplitude DSCT stars
Slow irregular variables	L	LB	late type giants
		LC	late type supergiants
Mira stars	M		long period late type giants
PV Telescopii	PVTEL		helium supergiant Bp stars
RR Lyrae	RR	RR(B)	double mode RR Lyr stars
		RRAB	RR Lyr stars with asymmetric light curves
		RRC	RR Lyr stars with symmetric light curves
RV Tauri stars	RV	RVA	radially pulsating supergiants with constant mean magnitude
		RVB	radially pulsating supergiants with variable mean magnitude
Semi-regular variables	SR	SRA	M, C, S giants with some periodicity
		SRB	M, C, S giants without periodicity
		SRC	
		SRD	
SX Phoenicis stars	SX PHE		pop II pulsating subdwarfs
ZZ Ceti stars	ZZ	ZZA	hydrogen pulsating white dwarfs
		ZZB	helium pulsating white dwarfs

Regarding the β Cep stars, *GCVS* has an entry BCEPS standing for 'short period group of β Cephei variables' and refers to the 'ultra-short-period B variables' (Jakate 1979), a group of which the existence is questioned. The *GCVS* definition of β Cephei variables embraces 53 Persei stars (see Section 3.4) which show light variations, and puts the upper limit of the period at $0^{d}.6$, instead of $0^{d}.3$ (see, for example, Sterken & Jerzykiewicz 1994).

Regarding the Cepheids (CEP in *GCVS*), we refer here to M.W. Feast in Section 3.7: 'The term Cepheid was at one time applied generally to any continuously varying star with a regular light curve and a period less than about 35 days, unless it was known to be an eclipsing star'. In Section 3.8, he also warns of the fact that a variety of designations have been used for the class of Type II Cepheids (CW stars) in the past, and also points out that, in practice, it may be difficult to unambiguously classify an individual star as a Type II Cepheid, an anomalous Cepheid, or a classical Cepheid, on the basis of the light curve alone, and that other information (e.g., galactic position, radial velocity, luminosity, chemical composition) is often used together with the light curve to classify Cepheids.

Lipunova (1990) mentions a specific new type of variable in the 5th volume of *GCVS*, viz. the BL Boo variables. First discovered by N.E. Kurochkin in 1961, it is now known to be an anomalous Cepheid having too great a luminosity for its period (or too small a period for its luminosity) – even for the continuation of the Population I period–luminosity relation. These stars are found, in particular, in dwarf spheroidal galaxies.

α Cyg variables (ACYG in *GCVS*) may exhibit eruptions, and often become Luminous Blue Variables (LBVs). Therefore, Section 3.1 should be read with parallel consultation of Section 2.1.

A few RR Lyrae stars show variations in the shape and amplitude of their light curves from cycle to cycle due to the simultaneous excitation of two periods (see Section 3.6). The *GCVS* refers to these double-mode RR Lyraes as RRb variables, though in recent literature these variables are often called RRd stars. M.W. Feast, in Section 3.8, makes an important point on the confusing nomenclature of classification of type II Cepheids in the 1 to 3 day period range, where the notation RRd is also used (though not in a consistent way).

In the *GCVS* the subtypes RVa and RVb for RV Tau stars are rendered as RVA and RVB. This usage is not recommended, because of possible confusion with the RV Tau subtypes A and B that are recognised on the basis of their spectra (see footnote by P.A. Whitelock in Section 3.9).

For irregular variables (slowly varying with no evidence of periodicity), the *GCVS* uses the classifications Lb and Lc for giants and supergiants, respectively. P.A. Whitelock, in Section 3.10, stresses that variables are often assigned to this class when their variability has been noted, but not well-studied: given that some semi-regular variables go through phases of irregular variations, it is not clear that the L classification represents a fundamentally different type of variability.

III. Rotating variables

GCVS: **'We call rotating variables the stars with non-uniform surface brightness or ellipsoidal shape, their variability being caused by their axial rotation with respect to an observer. The non-uniformity of surface brightness distributions may be caused by the presence of spots or by some thermal or chemical inhomogeneity of stellar atmospheres caused by magnetic field, its axis being not coincident with the star's rotation axis.'**

Table 1.4 Rotating variables

Variables	*GCVS*	
α^2 Canum Venaticorum	ACV &ACVO	B8p–A7p main-sequence stars
BY Draconis	BY	emission-line K–M dwarfs
Ellipsoidal	ELL	rotating ellipsoidal variables
FK Comae	FKCOM	rapidly-rotating spotted G–K giants
Pulsars	PSR	rapidly-rotating neutron stars
SX Arietis	SXARI	high-temperature analogues of α^2 CVn stars

Among the rotating variables, we discuss the Ap and roAp (rapidly rotating Ap stars), chemically peculiar stars (CP stars) where rotation plays a major role in the character of light variability. In the roAp stars (Section 4.1) though, the major cause of light variability is pulsation (combined with rotation). This only illustrates once more the difficulty of finding a consistent classification scheme for variables of truly polymorphic nature, even if a most objective and neutral approach is being followed. That following such an approach is not trivial, is most eloquently expressed by Don Kurtz while commenting on similarities between ZZ Ceti stars, roAp variables, and the Sun (Kurtz & Martinez 1994): 'Here we have three different kinds of pulsating star showing some remarkably similar behaviour. Yet the three different sets of astronomers working on these three different sets of stars have three different sets of explanations for the similar behaviour. That bears some thought.'

We do not discuss the helium variables (SX Ari variables, main-sequence B0p-B9p stars) which are the high-temperature analogs of the α^2 Canum Venaticorum stars.

IV. Cataclysmic (explosive and nova-like) variables

GCVS: **'We call explosive variables the stars showing outbursts caused by thermonuclear burst processes in their surface layers (Novae) or deep in their interiors (Supernovae).**

Table 1.5 Eruptive supernovae and cataclysmic variables

Variables	*GCVS*		
Novae	N	NA	Fast novae
		NB	Slow novae
		NC	Very slow novae
		NR	Recurrent novae
Novalike	NL		
Supernovae	SN	Type I Supernovae	SN I
		Type II Supernovae	SN II
U Geminorum stars (dwarf novae)	UG	UGSS	SS Cyg stars
		UGSU	SU UMa stars
		UGZ	Z Cam stars
Z Andromedae stars	ZAND	symbiotic systems	

We shall use the term 'Nova-like' for the variables showing nova-like outbursts due to rapid energy release in the surrounding space volume and also for the objects not displaying outbursts but resembling explosive variables in minimum light by their spectral (or other) characteristics.'

Some variables classified as nova-like in the *GCVS*, however, turn out to be LBVs or Be stars, see Section 2.1 and Section 3.3. On the other hand, *GCVS* does not mention the U Gem stars (Dwarf Novae, Section 5.4), which – even more than nova-like objects – show 'rapid energy release in the surrounding space volume'.

V. Eclipsing variables

GCVS: **'We adopt a triple system of classifying eclipsing binary systems: according to the shape of the combined light curve, as well as to physical and evolutionary characteristics of their components.'**

D.S. Hall, in Section 6.2, rightly points out that some binaries classified EB are not eclipsing at all: the light variation is produced entirely by an ellipticity effect (see Section 4.2) and the two minima are unequal as a result of greater limb-darkening effects on the pointed end of the highly distorted star.

Another type to draw attention to are the W Ser stars. The W Serpentis group was defined by Plavec (1980), but is not part of the *GCVS* classification scheme, and they can be described as a group of long-period Algol-like mass-transferring binaries, which are characterised by very substantial discs around the more massive components, strange and poorly-repeating light

Table 1.6 Eclipsing variables

Variables	GCVS			
Classification *a*	E	EA		Algol types
		EB		β Lyr types
		EW		WUMa types
Classification *b*		GS		one or two giant components
		PN		one component is the nucleus of a planetary nebula
		RS		RS CVn system
		WD		systems with a white dwarf component
		WR		systems with a WR component
Classification *c*		AR		AR Lac type detached system
		D	DM	Detached main sequence systems
			DS	Detached systems with subgiant
			DW	Detached systems like WUMa systems
		K	KE	contact systems of early spectral type
			KW	contact systems of late spectral type
		SD		semi-detached systems

Table 1.7 X-ray sources

Variables	GCVS		
X-ray sources	X	XB	X-ray bursters
		XF	fluctuating X-ray systems
		XI	X-ray irregulars
		XJ	X-ray binaries with relativistic jets
		XND	X-ray novalike with late-type component
		XNG	X-ray novalike with early-type component
		XP	X-ray pulsar system
		XPR	X-ray pulsar system plus reflection effect
		XPRM	X-ray system with late-type dwarf and pulsar with strong magnetic field

curves, prominent optical emission lines and large secular period changes (Wilson 1989) – typical members are RX Cas, SX Cas, W Ser, W Cru and β Lyr, and are classified in *GCVS* as EA/GS or EB/GS (see also Andersen *et al.* 1988).

VI. X-ray sources

GCVS: **'Close binary systems, sources of strong variable X-ray radiation, which do not belong to any of the above types of variable stars.'**

The nomenclature on X-ray sources is quite consistent, however, we refer to J. Krautter's remark in Section 4.5 who points out that 'One may not confuse radiopulsars with X-ray pulsars which emit pulses in the X-ray regime; they belong to the group of X-ray binaries'.

1.4 Bibliography of variable stars

H.W. Duerbeck

The growing number of variable stars and the growing number of publications containing observations of a single variable star, or an interpretation of the observed phenomenon, or publications mentioning a variable star among others, from the very beginning led to bibliographical compilations. Changing trends in the type of publication are obvious: while the first bibliographies appeared in printed form, the next generation was distributed as microfiche, or even on magnetic tape, while the most recent compilation of references is available by anonymous *ftp* from an astronomical data centre. The printed bibliographies explicitly give references to positions, charts, comparison charts, light curves and spectroscopy, and the most recent machine-readable bibliographies by Roessiger & Braeuer (1992, 1993, 1994, 1995) and Roessiger (1992) give information on the contents of listed publications by a system of about one hundred symbols. Such information is, however, lacking in the bibliography of Huth & Wenzel (1981, 1992), so that the user has to 'dig for him/herself'.

We would like to give a 'historical' overview of the bibliographies, because the more recent ones do not repeat the information which is contained in the older (printed) versions. In Table 1.8 we list the printed bibliographies, published under the general title *History and bibliography of the light variations of variable stars* (text usually in German), and also the new machine-readable ones.

As can be seen, the location of information for a given variable star is not obvious from Table 1.8, because the time of discovery and naming is often unknown. However, if the user has access to a computer connected to a network, the most convenient procedure to obtain bibliographic information on a variable star is to consult electronically the SIMBAD database. Telnet to `simbad.u-strasbg.fr` – the user has to apply for a user account – and enter the variable star designation (e.g. `V* RR LYR, V* DELTA CEP, V* V616 MON`). This database also offers a listing of the bibliography (from about

Table 1.8 Overview of bibliographies on variable stars (with Edition, Volume and Issue)

Author and Year	Ed.	Vol.	Iss.	Contents
Müller & Hartwig 1918	I	1		literature of stars named until the end of 1915, R.A. 0 - 14 hrs
Müller & Hartwig 1920	I	2		literature of stars named until the end of 1915, R.A. 15 - 23 hrs, plus novae and variables in clusters
Müller & Hartwig 1922	I	3		elements, stars named 1915-1920, novae
Prager 1934	II	1		literature 1916-1933 for stars named before 1930 in And-Cru
Prager 1936	II	2		literature 1916-1933 for stars named before 1930 in Cyg-Oph
Prager 1941	II	suppl.		literature for stars named 1931-1938
Schneller 1952	II	3		literature 1930-1950 for stars named before 1938
Schneller 1957	II	4		additional literature until 1954 for stars named before 1938
Schneller 1960	II	5	1	stars named 1938-1958 in And-Cyg
Schneller 1961	II	5	2	stars named 1938-1958 in Del-Per
Schneller 1963	II	5	3	stars named 1938-1958 in Phe-Vul
Huth & Wenzel 1981, 1992	-	-	-	Bibliographic Catalogue of Variable Stars Part I (Literature to 1982)
Roessiger & Braeuer 1992, 1993	-	-	-	Bibliographic Catalogue of Suspected Variable Stars Update (Revised)
Roessiger & Braeuer 1994, 1995	-	-	-	Bibliographic Catalogue of Variable Stars Part II (Literature 1982–1994)

1950 onwards) in major astronomical journals. However, if a more complete bibliography (including amateur journals, observatory publications, books and catalogues) is required, the bibliographic catalogues of Huth & Wenzel and Roessiger & Braeuer (cats/VI/67 and cats/VI/68) can be downloaded via *ftp* from `cdsarc.u-strasbg.fr`. After uncompressing, the interesting sections can be copied e.g. with the unix command: `egrep 'LYR RR' main67 > filename` (or, for the other two examples, `'CEP DEL'` `'MON 616'`) (the files are too large to be accessed easily by a text editor). These bibliographic catalogues also give hints on which parts of the printed bibliographies of Prager, Müller & Hartwig, and Schneller have to be consulted. It is trivial to note that all these bibliographies have deficiencies.

Finally, we should mention some additional bibliographies: the *Bibliographic Catalogue of Suspected Variable Stars* (*BCSVS*, Roessiger & Braeuer 1993, 1994), available via ftp from `cdsarc.u-strasbg.fr` as cats/VI/58, contains the bibliographic references of stars listed in the *New Catalogue of Suspected Variable Stars* (*NSV*, Kukarkin et al. 1982). For observers of eclipsing variable stars, the *Finding List for Observers of Interacting Binary Stars* (Wood et al. 1980) supplies a good overview of all eclipsing systems, sorted according to right ascension, with short bibliographical notes.

1.5 Variable stars, photometric systems and photometric precision

C. Sterken

Throughout this book we refer, in general, to photometric measurements obtained differentially – that is, using a nearby suitable comparison star of the same colour and of similar magnitude as the variable star. In order to facilitate literature searches, we have included (when possible) the name of the comparison star in the figure captions, and also in the Object Index.

The data used are, in general, either obtained in the Strömgren *uvby* system, the Walraven *VBLUW* system, the Geneva $UBVB_1B_2V_1G$, or the well-known *UBV* system (note that the *UBV* pass bands of the latter systems are similar, though not identical). Thus, when we use *uvby*, *VBLUW* or *UBV* light curves, we refer to data coming from one of these main systems.[1] Therefore, our graphs should be considered as *qualitative* illustrations, and should in no way be combined or merged to produce a single, well-covered diagram. Instead, the reader should go back to the original sources of data. In doing this, users of data catalogues should carefully read the introductory texts to catalogues, and not only take the listed data at their face value. Many a user of the Geneva photometric catalogue (Rufener 1988) does not immediately realise that the listed U, B_1, B_2, V_1, G magnitudes are colour indices (they are relative to the measured B), and not magnitudes, and it may take considerable time to understand why a star that is variable in b or y, may – at the same time – be perfectly constant in V_1. And it is often forgotten that the Walraven *VBLUW* magnitudes and colour indices are always expressed in $\log I$, and not in magnitudes, which makes a difference by a factor −2.5. We refer to Figs. 1.4–1.6 for an idea of the shape and position of the pass bands of the $uvby, UBV/UBVB_1B_2V_1G$ and *VBLUW* systems.

The study of the photometric variability of stars is intimately related to the precision and accuracy of photometric magnitude and colour measurements. Variable-star observers, through history, contributed a great deal towards the improvement of photometric techniques and to the establishment of adequate photometric standards. The literature often gives impressively small formal errors on photometric measurements, but it is rather difficult to assess the real uncertainty of the results, and to assign a reliable error bar to absolute

1. However, some *UBV* light curves are also taken from data collected in the Cousins *UBVRI* system.

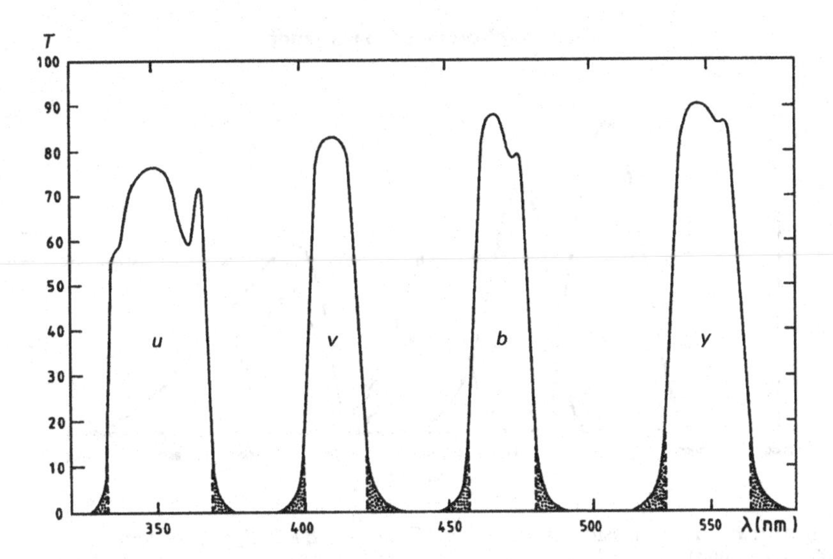

Figure 1.4 Transmission curves of the Strömgren *uvby* system (Sterken & Manfroid 1992, see also Florentin Nielsen 1983). The shaded areas indicate the wings cut off by the slots of the spectrometer.

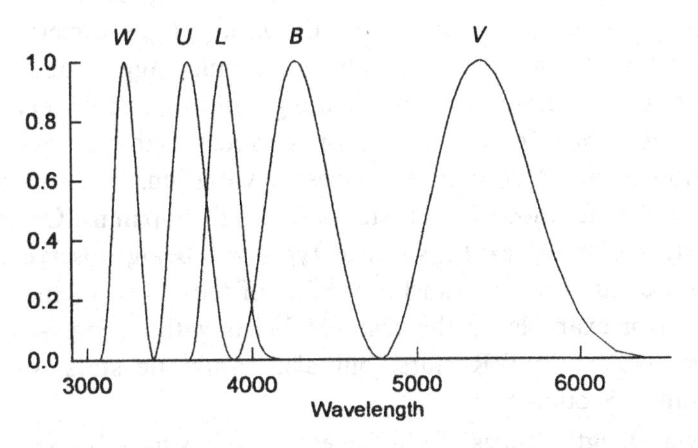

Figure 1.5 Normalised transmission curves of the Walraven *VBLUW* system (Lub & Pel 1977).

photometric data. Many users of a photometer are not aware of the existence of systematic errors that are introduced during the photometric reduction process. Even when obtaining a fairly good internal precision, one often has no idea what the real accuracy of a measurement is. There are many reasons for this: the most important ones are the problems of atmospheric extinction correction, and the complications associated with the transformation of the data from a local instrumental system to a universal standard system, and these steps depend on

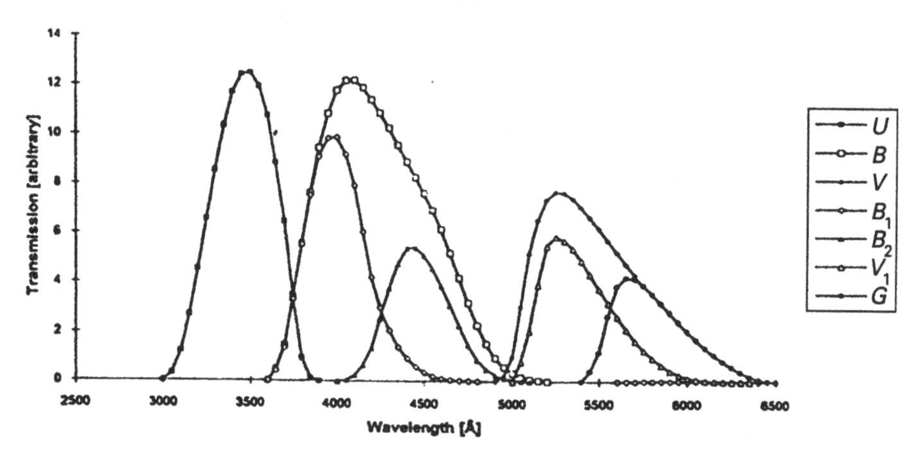

Figure 1.6 Transmission curves of the Geneva $UBVB_1B_2V_1G$ system (Cramer 1994).

many aspects of instrumental conformity and of methodological rigorousness. Problems with standardisation are as old as the craft of photometry. But for variable star observing, the existing problems are much aggravated by the ever present restriction of phase coverage, leading too often to observing at excessively high air masses. Moreover, for short-period variables, there is the need for observation of many contiguous cycles of variation, in turn leading to instrumental incompatibilities in multi-site observing campaigns. On top of all this, variable stars of most exotic spectral types are being observed, and the most elementary conditions for transformability of these measurements are constantly violated (for example, in the case of objects with strongly-variable emission lines, like LBV and WR stars, but also symbiotic stars, see P.A. Whitelock's warning in Section 5.5).

In the past the wavelength ranges of photometric systems have been dictated by the means available rather than by the astrophysical problem to be studied. But today, with most photometric systems being established, the application of a specific photometric system to a particular type of variable star is, again, mainly dictated by the availibility of the photometric system at the observing facility where the data are collected. Sometimes, a specific photometric system is available on a given site, but the photometer is mounted on a telescope too small to yield a favourable photon budget, forcing the observer to work with a less-optimal instrumental configuration.

Figure 1.7, taken from Sterken (1992) is a graphical interpretation of the table of ancestry of existing photometric systems (only the well-known systems

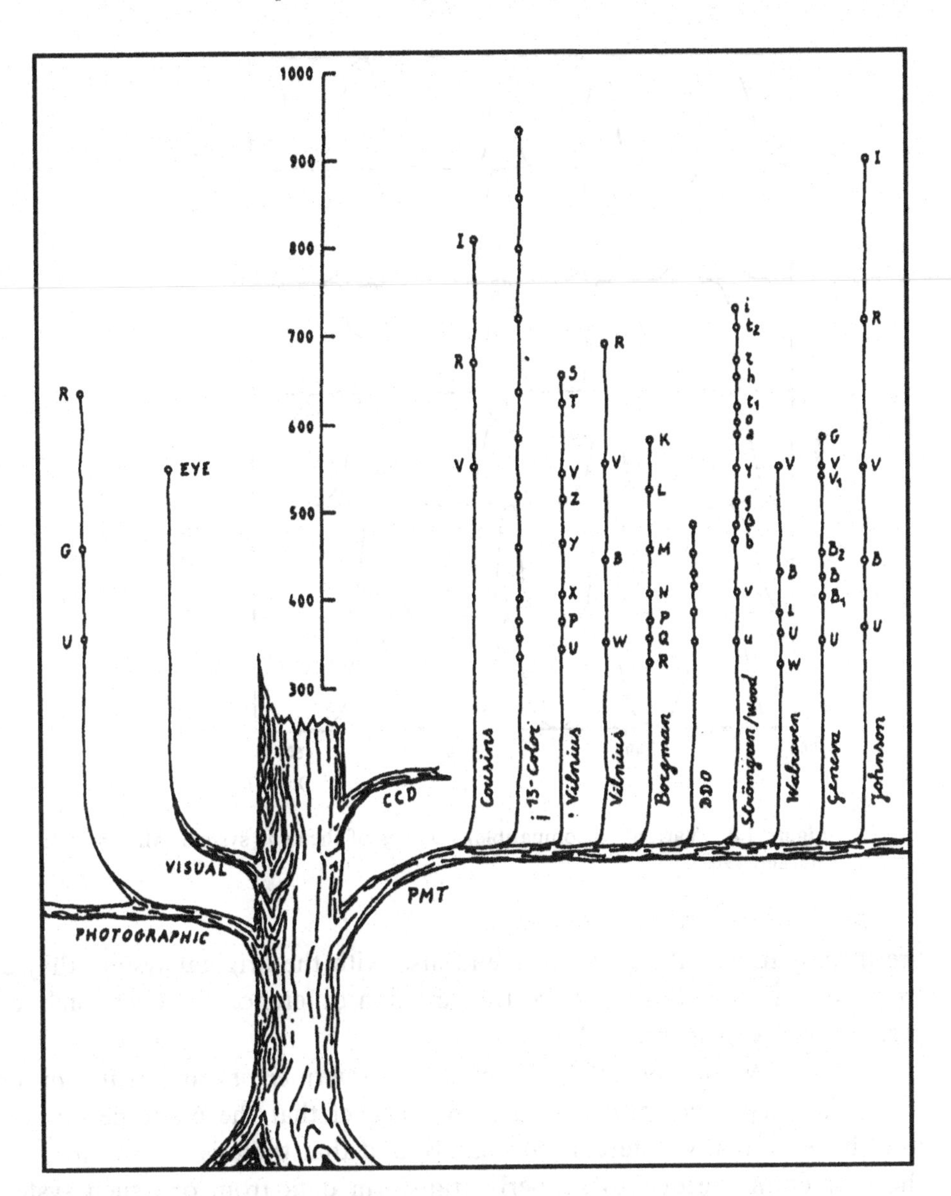

Figure 1.7 Genealogy for some well-known photometric systems. Open circles indicate the position of the central wavelength of the indicated systems (in nm, increasing along the vertical direction). (Sterken 1992).

are indicated). The four main branches (visual, photographic, photomultiplier tube and CCD) are detector branches, from which the individual systems have grown. Note that this representation is an idealised situation in the sense that it depicts the view as seen by the designer of each system. It does not show the reality in which several 'pure' systems are accompanied by 'clones' which

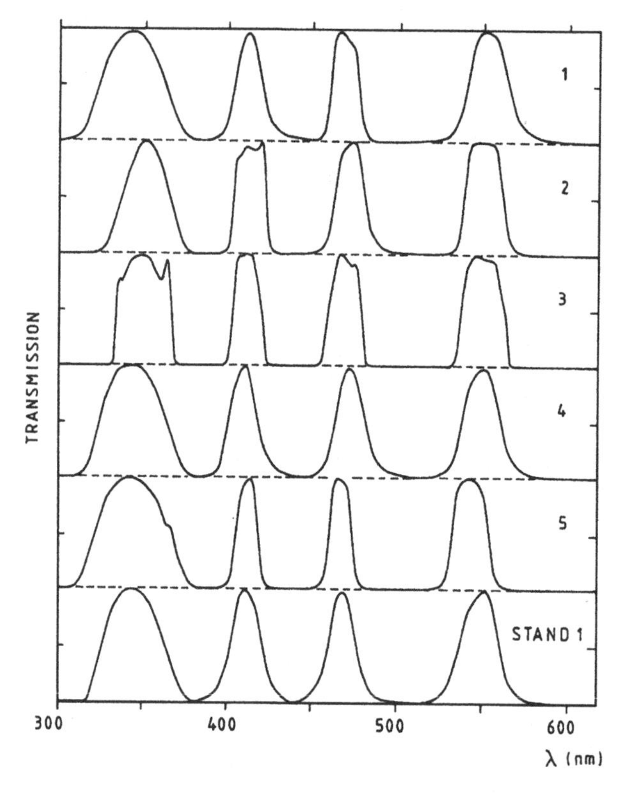

Figure 1.8 Various (incompatible) versions of the *uvby* system (Manfroid & Sterken 1987).

are incompatible with each other and also with the original system they came from. For an illustration of the transmission differences between such 'clone' systems, we refer to Fig. 1.8.

Manfroid & Sterken (1992) discern *conformity errors* and *reduction errors* in photometry. The former arise from the fact that the photometric systems used have mutually different pass bands, and that there is no way to evaluate the corrections needed to properly transform data from one such system to another. The latter are of a purely methodological nature and are due to the reduction procedure. As said already, conformity errors are often unavoidable, since prescriptions of a purely practical origin (such as the availability of a given photometric system at one telescope) may force the investigators to rely on data coming from different systems. Reduction errors can arise between batches of data reduced separately, as is the case in long-term and network projects. Some of the parameters in the reduction schemes have larger errors than others, and these uncertainties on the colour coefficients translate into errors in the colour indices themselves; the resulting errors are appreciably

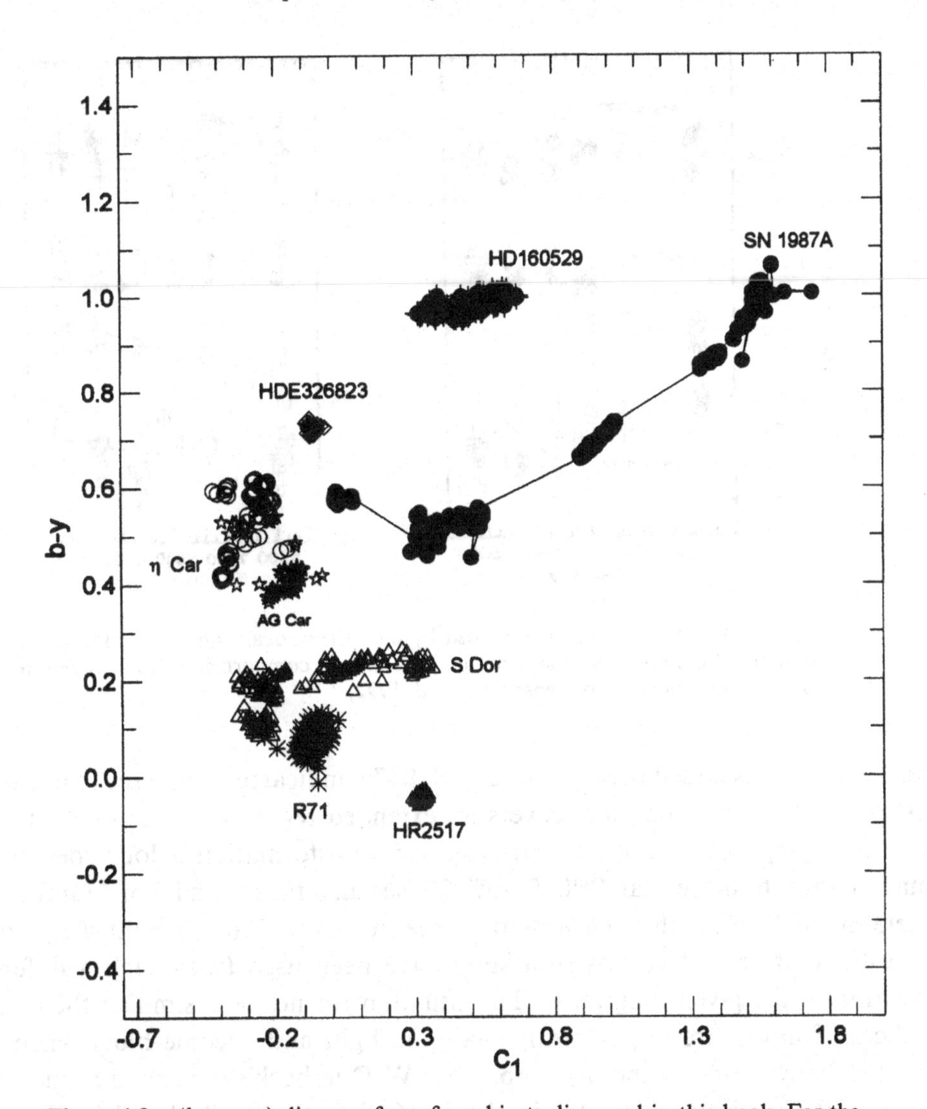

Figure 1.9 $(b-y, c_1)$ diagram for a few objects discussed in this book. For the sake of clarity, consecutive points for SN1987a have been connected by a line. The diagram is based on *uvby* data taken from Manfroid *et al.* (1991, 1994) and Sterken *et al.* (1993, 1995b).

larger for stars with extreme colour indices. Such effects are random shifts that affect all measurements of a given star by the same amount (during a specific observing run). For a more detailed account, see Sterken (1993).

Figure 1.9 shows a $(b-y, c_1)$[1] diagram for several objects discussed in this book.

HD 160529 is a a highly-reddened LBV located in the direction of the galactic

1. $c_1 = (u-v) - (v-b)$ is a Strömgren photometric index; for early-type stars the c_1 index is an effective-temperature index.

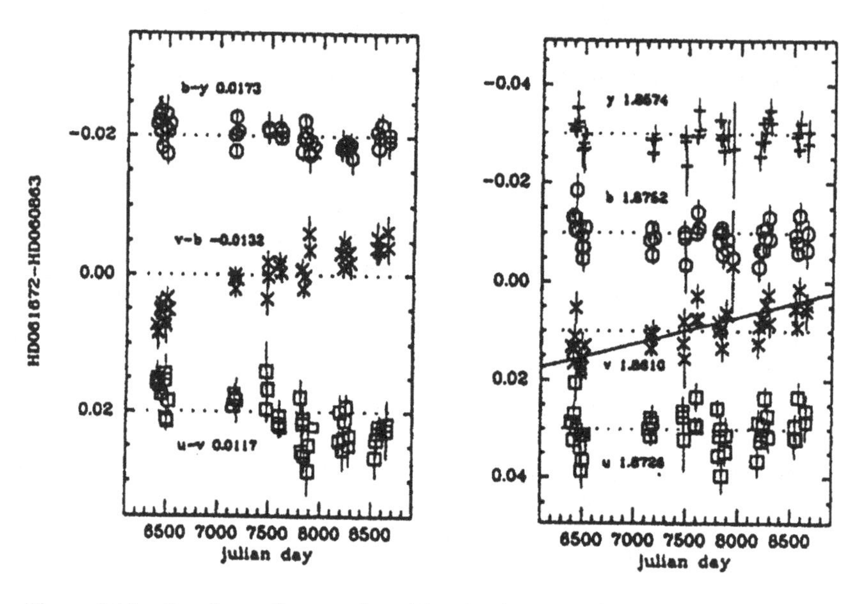

Figure 1.10 Spurious effect produced by the drift in dead-time correction for the v-channel: differential photometry for two constant comparison stars of different brightness (adapted from Hensberge *et al.* 1992).

center, and is discussed in Section 2.1. SN1987a, indicated by crosses to illustrate its long path in the diagram, covers an extended region in the $(b-y, c_1)$ (and also $b-y, m_1)$ space, and the errors on the transformation colour coefficients may amount to more than $0^{\text{m}}.015-0^{\text{m}}.020$ (see also Figs. 2 and 3 of Manfroid & Sterken 1992). Note that, because of the large changes in apparent magnitude, two different sets of comparison stars have been used (with each a different position in the given diagrams). The situation for novae is similar: the colour indices of novae are largest near maximum light and become bluer when the photosphere contracts and heats up, see H.W. Duerbeck's remark in Section 5.2. R 81, a P Cygni-duplicate star in the LMC, is rather well-placed in Fig. 1.9; remark, though, the large range of variation in c_1.

The CP stars HD 61672 and HD 201601 (γ Equ) provide an illustrative example of the significant interpretational consequences of subtle errors. As debated above, differing instrumental versions of a photometric system, non-congruent transformation schemes, but also orderly transformations to a standard system, occasionally may introduce deceptive details in the shapes of the light curves of variable stars. Such effects are most pronounced in those cases where the light variability is wavelength-dependent (like in the case of CP stars, Section 4.1), because the transformation equations (see, for example, Young 1992) have no relation with the physical processes underlying the wavelength-dependency of the light variations. In such cases, differential photometry in different magni-

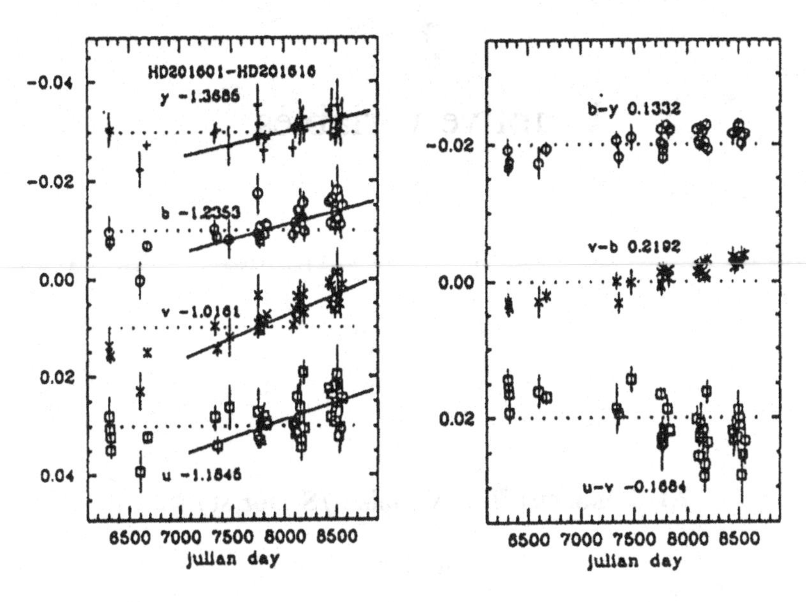

Figure 1.11 Real magnitude and colour changes of γ Equ (HD 201601) relative to HD 201616. The vertical scale is a relative scale; the mean magnitude and colour-index values are indicated (adapted from Hensberge *et al.* 1992).

tude systems, even photometry obtained with a single version of a particular photometric system (but transformed to a standard system), is inappropriate. Hensberge (1993) discovered an instrumental trend of long-term character at the level of the dead-time correction associated with the *v* passband (*uvby* system at the SAT telescope at ESO): *v*-data (and colour indices involving the *v* passband) of couples of comparison stars of considerably differing magnitudes showed systematic drifts, and the effect was confirmed with non-differential results on bright stars afterwards (see Fig. 1.10). The nominal value for the dead time, 8.8 10^{-8} s, seemed to change at a rate of 1.7 10^{-8} s per 1000 days, resulting in a change of about 35% over five years. Figure 1.11 shows real light variations of γ Equulei, the slowest-rotating CP star known (once per century, Mathys 1990), and clearly illustrates that only very careful analysis can make us discern between real and spurious variations of such small amplitude.

It is clear that the above-mentioned elements contributing to error and uncertainty may lead to a situation where small changes in the instrumental system (like the change of a photomultiplier), small changes in the electronic counting circuitry (dead-time constant), small transformation errors (due to the design properties of the photometric system), and small reduction errors (due to the limited sizes of batches of data under reduction, but also because of the presence of large-amplitude variability) may mimic, change, or even conceal small intrinsic variations.

2

Eruptive variables

2.1 Luminous Blue Variables/S Dor stars

C. Sterken

The observed H–R diagram has an upper luminosity limit of which the contour line is temperature dependent (Humphreys & Davidson 1987). Some of the most massive and luminous ($10^6 L_\odot$) stars near that line (P Cyg, AG Car, HR Car, η Car, ...) sporadically show dramatic mass-ejections (seen as 'eruptions') followed by periods of quiescence. Such stars are called hypergiants, some of them are Luminous Blue Variables[1] (LBVs), though LBVs do not necessarily need to be blue, since the phenomenon is not restricted to early-type stars (de Jager & van Genderen 1989). The above-mentioned LBVs, together with 164 G Sco = HD 160529 (Fig. 2.1) and WRA 751 are notorious galactic LBVs. de Koter (1993) estimates the number of LBVs in our galaxy at no more than 60, but the number of LBVs that possibly can be considered for observation, obviously, is much less. In the LMC, the well-known LBVs (also called S Dor stars) are S Dor (Fig. 2.2), R 71 (Figs. 2.3, 2.4) and R 127 (Fig. 2.5), with R 66, R 81 (Fig. 2.6) and R 110 as additional candidates. Finally there are the 'Hubble–Sandage variables', discovered by Hubble & Sandage (1953) in M31 and M33, which are identifiable with the S Dor variables, that complete the group that is commonly designated as LBVs. During outburst these stars are – apart from supernovae – the visually brightest stars in the universe, and thus potentially belong to the most powerful extragalactic distance-indicators (Wolf 1989). Today, only a few dozen LBVs are known.

P Cyg is a most notorious example of an LBV: it first became visible in 1600, was highly variable (with irregular fadings to below the $V = 6$ limit, see

1. See van Genderen (1979) and Conti (1984) for the introduction of this nomenclature.

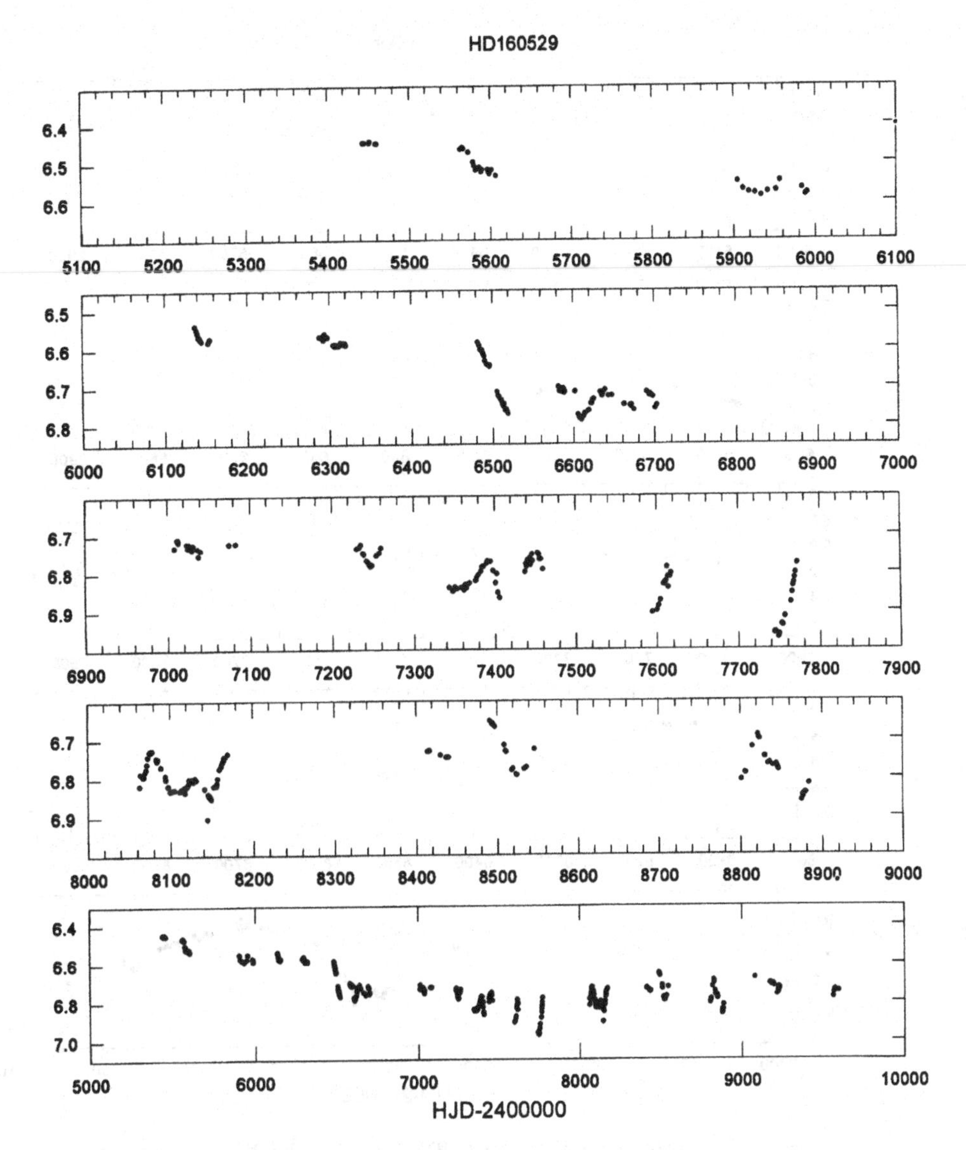

Figure 2.1 *V* light curve of HD 160529 (LBV) (based on LTPV data, see also Sterken *et al.* 1991). The lower panel shows the very-long-term behaviour.

Fig. 2.7) and now it is of about fifth magnitude (for more details, see Fig. 2.8 and de Groot 1969, 1988).

Another famous example of a galactic LBV is η Car (formerly called η Argus), one of the most enigmatic objects in the Milky Way, it is also one of the most extreme infrared sources in the sky (Neugebauer & Westphal 1968). The light variability of η Car had already been described by Innes (1903) and by Gratton (1963). The star showed large light variations during the last century: being of

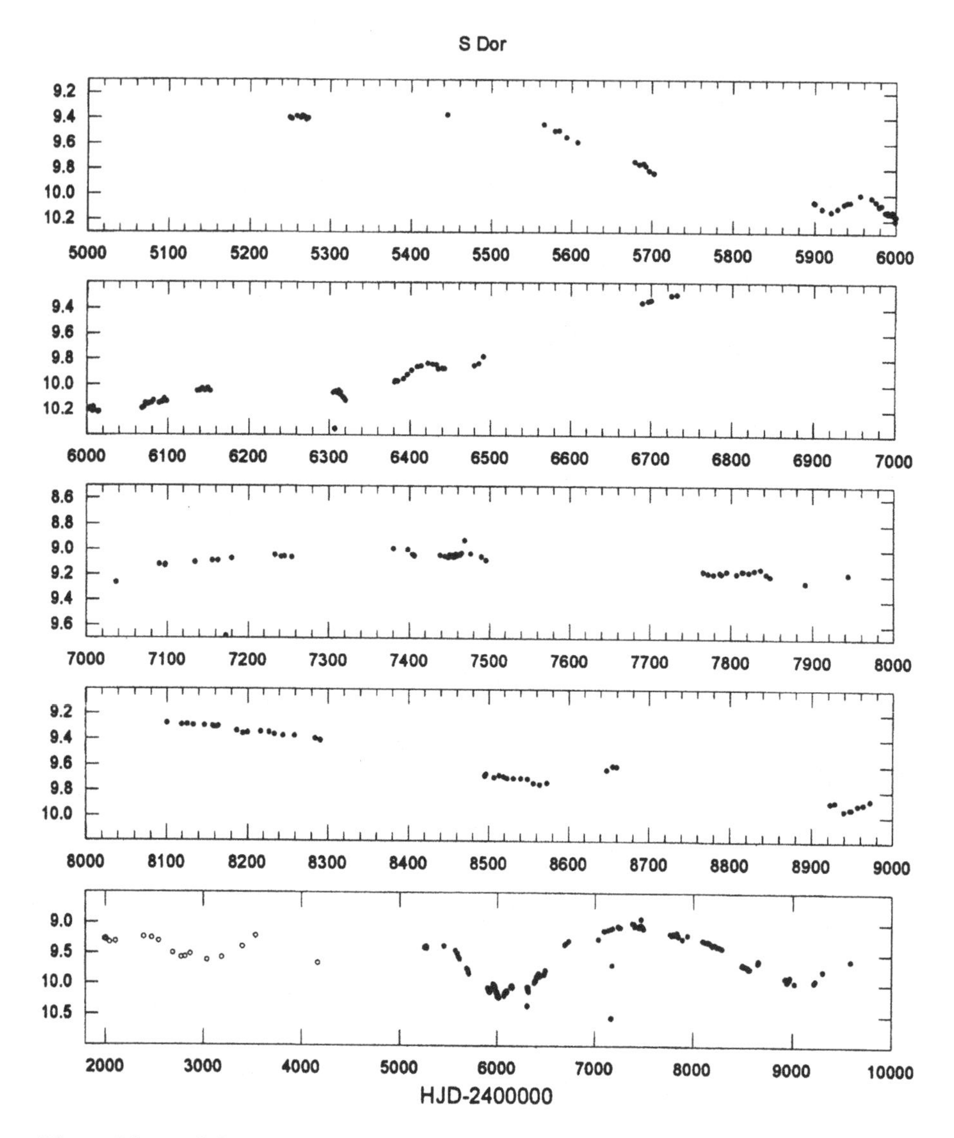

Figure 2.2 *V* light curve of S Dor (LBV) (based on LTPV data and data discussed by Spoon *et al.* 1994).

magnitude 2–4 at the beginning of the 1800s, it attained first magnitude in 1843 and was the second brightest star (after Sirius) in the sky (Viotti 1990). During the 14 years following a new short brightening, η Car underwent a spectacular light decrease from the first to seventh magnitude (a fading caused by formation of a dusty envelope, a very common characteristic of LBV variability).

Very detailed descriptions of the light variations (see, for example, Fig. 2.9) of η Car have been published by van Genderen & Thé (1984) and van Genderen

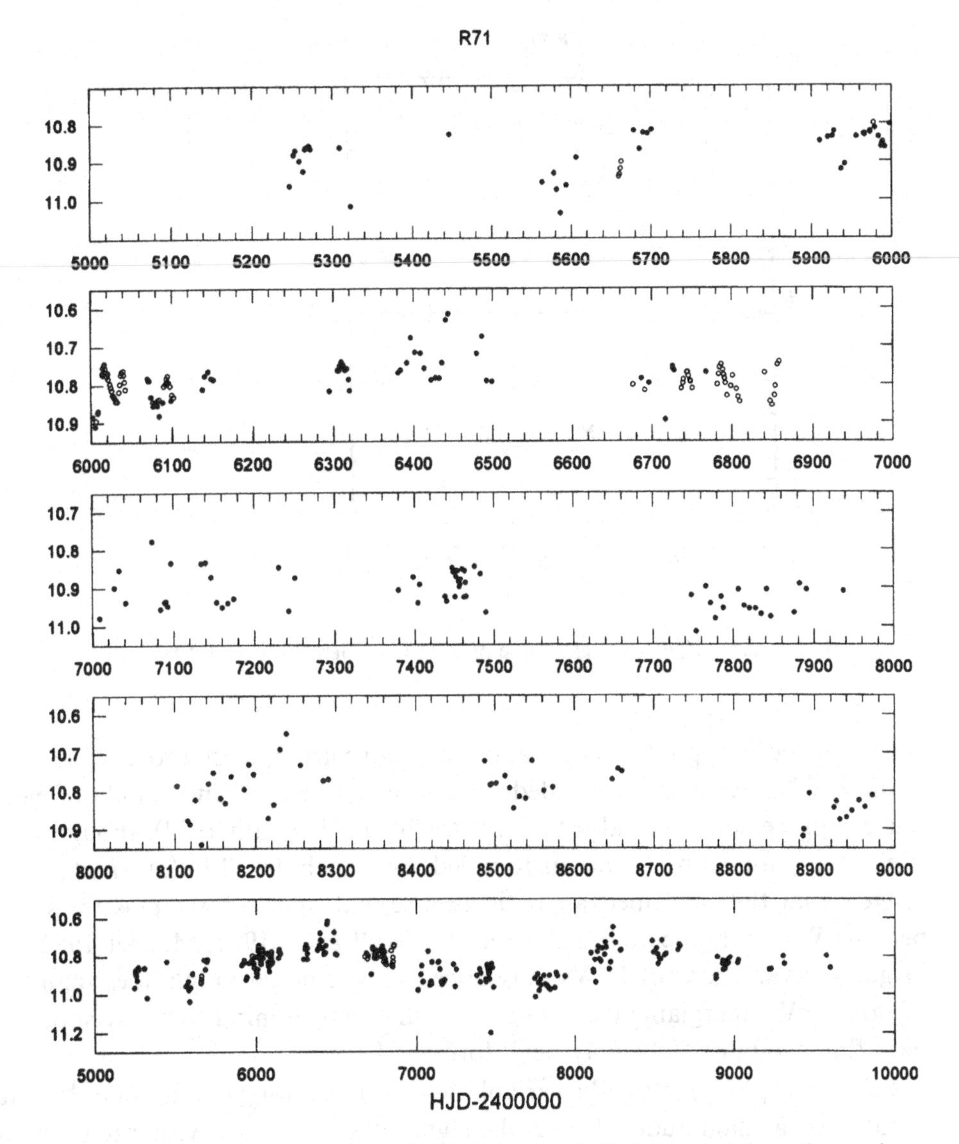

Figure 2.3 *V* light curve of R 71 (LBV) (based on LTPV data and data discussed by Spoon *et al.* 1994).

et al. (1994, 1995), see also Zanella *et al.* (1984). For information on the variability at infrared wavelengths, we refer to Whitelock *et al.* (1994), who found quasi-periodic variations on a time scale of about 5 years representing S Dor phases.

Most S Dor/LBV variables have a P Cygni type spectrum (strongest Balmer lines and HeI lines in the visual spectrum show P Cygni profiles – that is, an emission component with blue-shifted absorption component); actually the

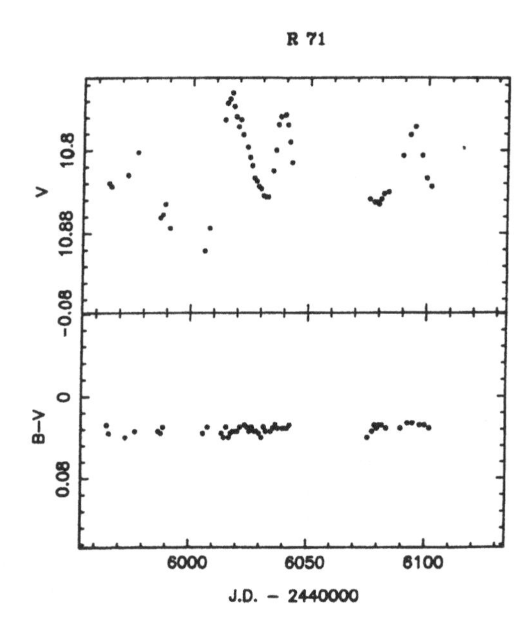

Figure 2.4 $V, B-V$ light curve of R 71 (van Genderen *et al.* 1985).

classes of the P Cygni type stars and the S Dor variables are probably identical. Some LBVs have a very extended circumstellar shell, or ring nebula of ejected nuclear-processed material (Stahl 1989, Chu 1991, Smith 1993), though P Cyg itself does not seem to be surrounded by a nebula. Moffat *et al.* (1989) suggest that the LBV mechanism is an essential step to force massive stars to become WR stars (Section 2.2) – see also Wolf *et al.* 1981, Maeder 1982). The brightness variations of LBVs in quiescence have larger amplitudes than those of non-LBV supergiants (though one should bear in mind that any supergiant near the instability limit may be a dormant LBV).

LBVs are photometrically variable with a large range of amplitudes (hundredths of a magnitude to several magnitudes) and on a vast range of time scales (hours, over several decades, to centuries). Humphreys & Davidson (1994) call LBVs 'Astrophysical Geysers with irregular eruptions of varying degree' recognised by 'moderate activity, then violent burbling, then the fitful eruption itself, followed by a period of relative quiescence'. Variability is partially due to rapid and unsteady mass loss, but several additional mechanisms have been proposed to account for the observed variations – one possibility is pulsational instability, since the β Cep instability strip (Section 3.2) widens into the supergiant region, an explanation that is supported by theoretical calculations (Dziembowski 1994) and by observational evidence (Sterken 1989, Sterken *et al.* 1995a). The amplitudes of the variations seem to increase with

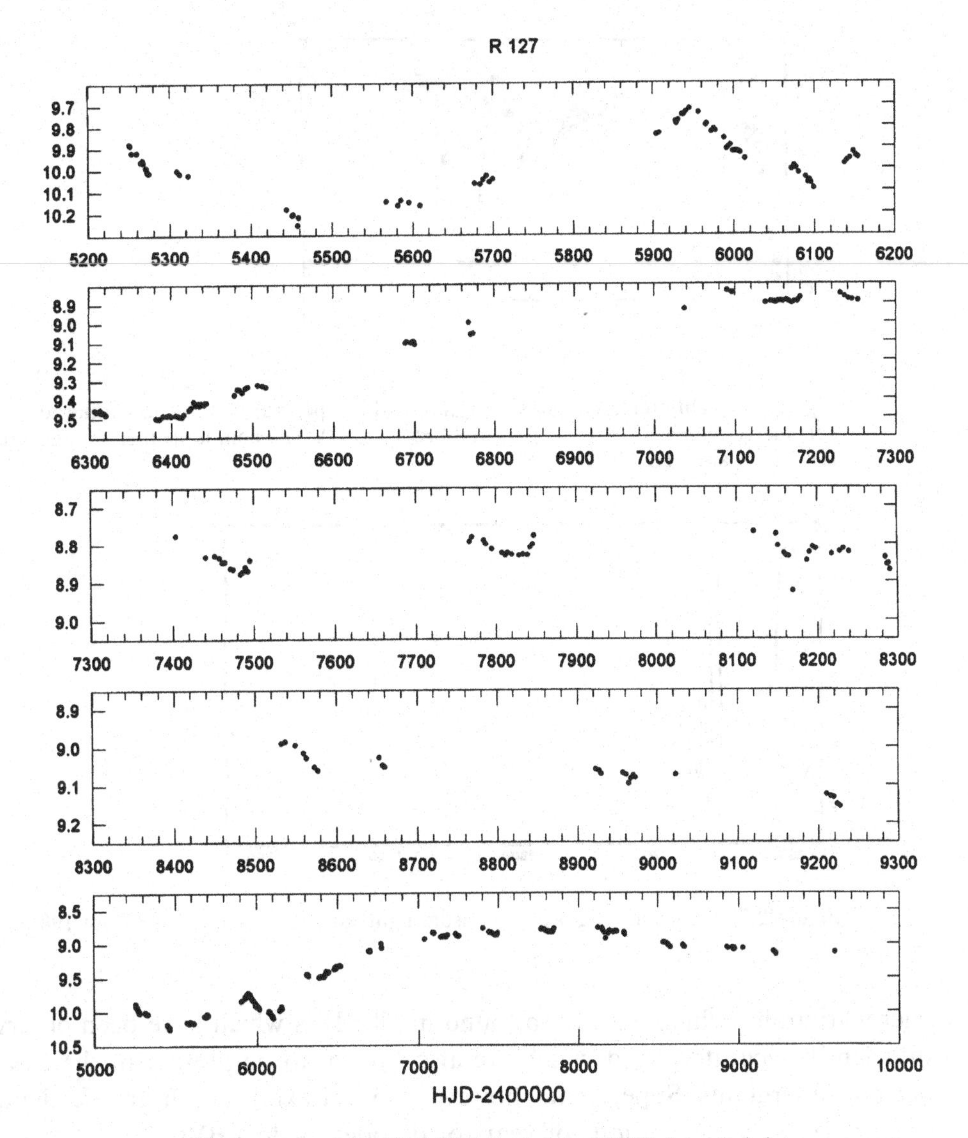

Figure 2.5 *V* light curve of R 127 (LBV) (based on LTPV data and data discussed by Spoon *et al.* 1994).

the time scales at which they occur. There are three types of variations: the large variations (which are associated with eruptions, and which are seen in some, but not in all LBVs, and which occur on time scales of centuries[1]), the moderate variations (seen on time scales of decades, and which occur at irregular intervals), and the small-scale photometric variations, also called

1. Such eruptions can be mistaken for nova or supernova events, others are of such magnitude that there are doubts whether they should be called eruptions, see Davidson (1989).

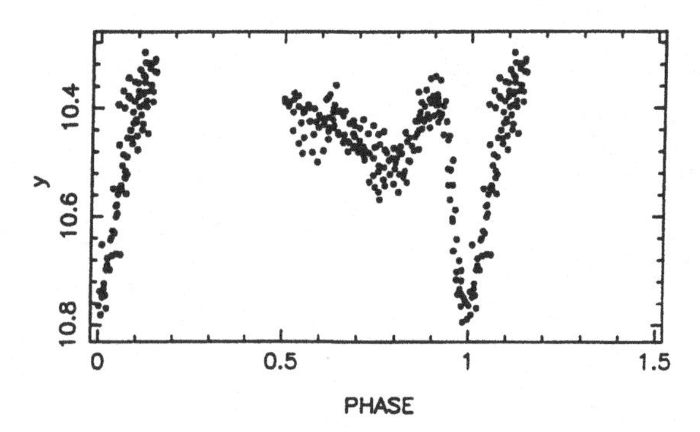

Figure 2.6 Differential y phase diagram (orbital period) of R 81 according to ephemeris 2445973.329 + 74.59E (Stahl *et al.* 1987). The comparison stars used are HD 34144 and HD 34651.

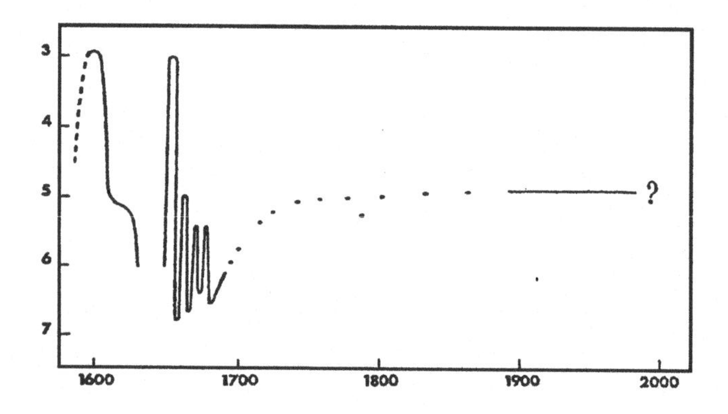

Figure 2.7 Schematic light curve (visual light) of P Cyg (LBV) (de Groot 1988).

microvariations, which have been found in all LBVs which have been observed sufficiently accurately, and which are also known to be present in the case of normal supergiants. Supergiants, however, do not exhibit such drastic changes of spectral type over periods of years or decades as do LBVs.

These microvariations are not strictly periodic, and the characteristic period is not always unambiguously visible. In the case of P Cyg, for example, the range of time scales goes from 25 days to 60 days (van Gent & Lamers 1986). These time scales are not constant in length. In the case of R 71 (Fig. 2.4) the microvariations are characterised by regular and smooth ups and downs with an amplitude of about $0\overset{m}{.}1$, and with a semi-period of about 23 days (in 1983–1985, van Genderen *et al.* 1985) and of about 14 days (in 1986–1987, van Genderen *et al.* 1988). The colours vary too, but the amplitudes of variation are a few orders of magnitude smaller than the light amplitudes. The colour curves are usually in phase with the light curves, and the light amplitudes tend

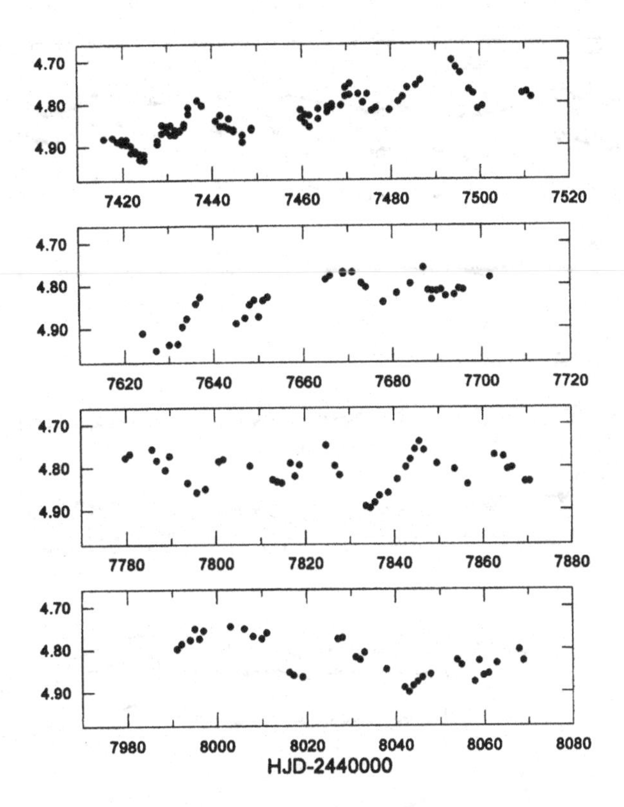

Figure 2.8 *V* light curves of P Cyg (LBV) based on data from M. de Groot obtained with a 10-inch APT.

to increase with decreasing wavelength. Stars similar to R 71 in relation to their variability, are AG Car (Fig. 2.10) and HR Car. The time interval between successive eruptions is of the order of 5 years, and the brightness rises seem to be characterised by a very steep rising branch, and by a descending branch with a gradient that is about 5 times less. The microvariations in the quiescent state have a characteristic time of about 10–14 days and 20 days, respectively, and there are large cycle-to-cycle shifts in the mean light level present due to weak S Dor phases of the same type as exhibited by P Cyg (van Genderen *et al.* 1988, 1990a).

The highlight among LBVs undoubtedly is R 81 (Fig. 2.6), an eclipsing binary, and the only one known in that class. From data collected over several years, it is evident that dips in the light curve with a depth of $0\overset{m}{.}4$ occur at regular intervals of about 75 days; there is no secondary minimum visible. The light curve indicates a contact system with complicated mass motions. Strangely, the microvariations of R 81 itself are unusually small for a supergiant.

Figure 2.11 shows light- and colour curves of HD 168607, an LBV according

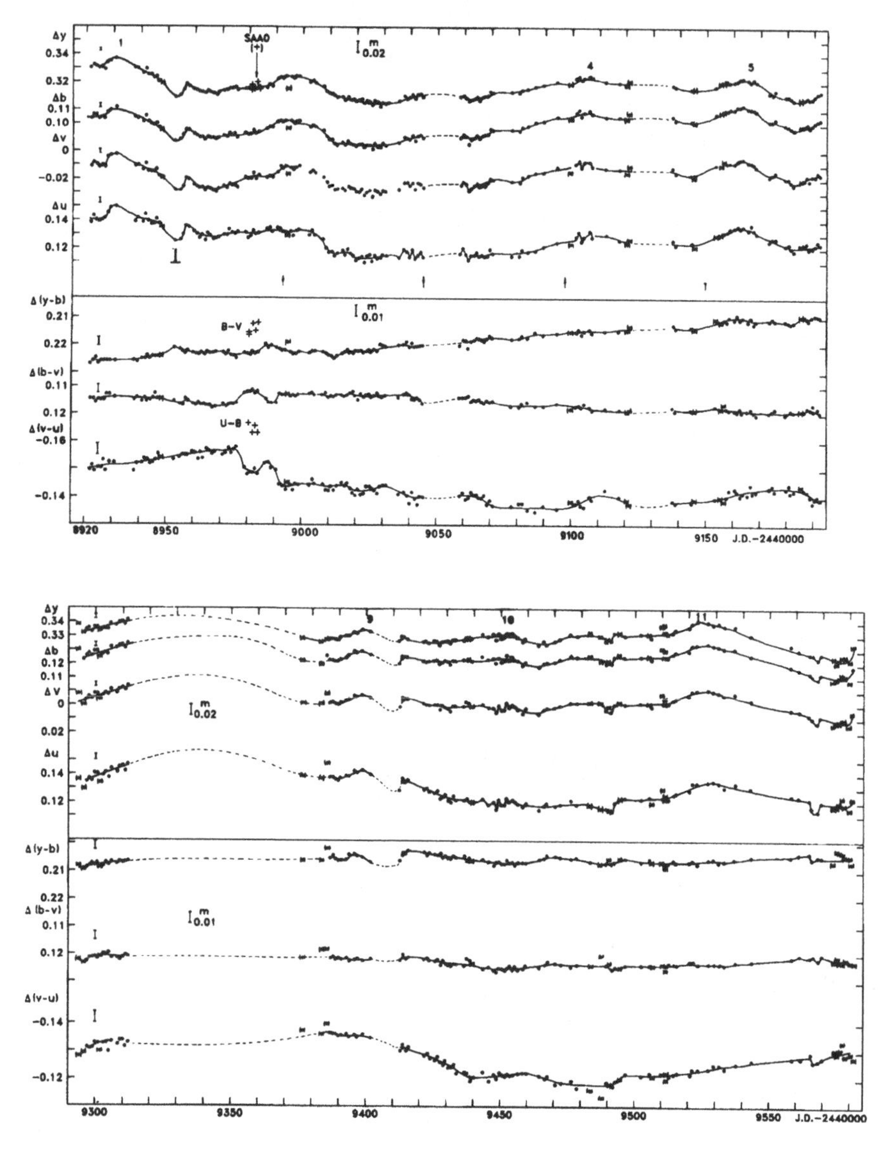

Figure 2.9 *uvby* light- and colour curves of η Car (LBV) (van Genderen *et al.* 1995).

to Chentsov & Luud (1990), and very well illustrates the LBV characteristics – that is, according to van Genderen *et al.* (1992)

(i) the total light range or maximum light amplitude (MLA, see van Genderen *et al.* 1992) is higher than for normal α Cyg variables of the same temperature
(ii) small range oscillations (time scale $\sim 60^{\mathrm{d}}$ occur on different brightness

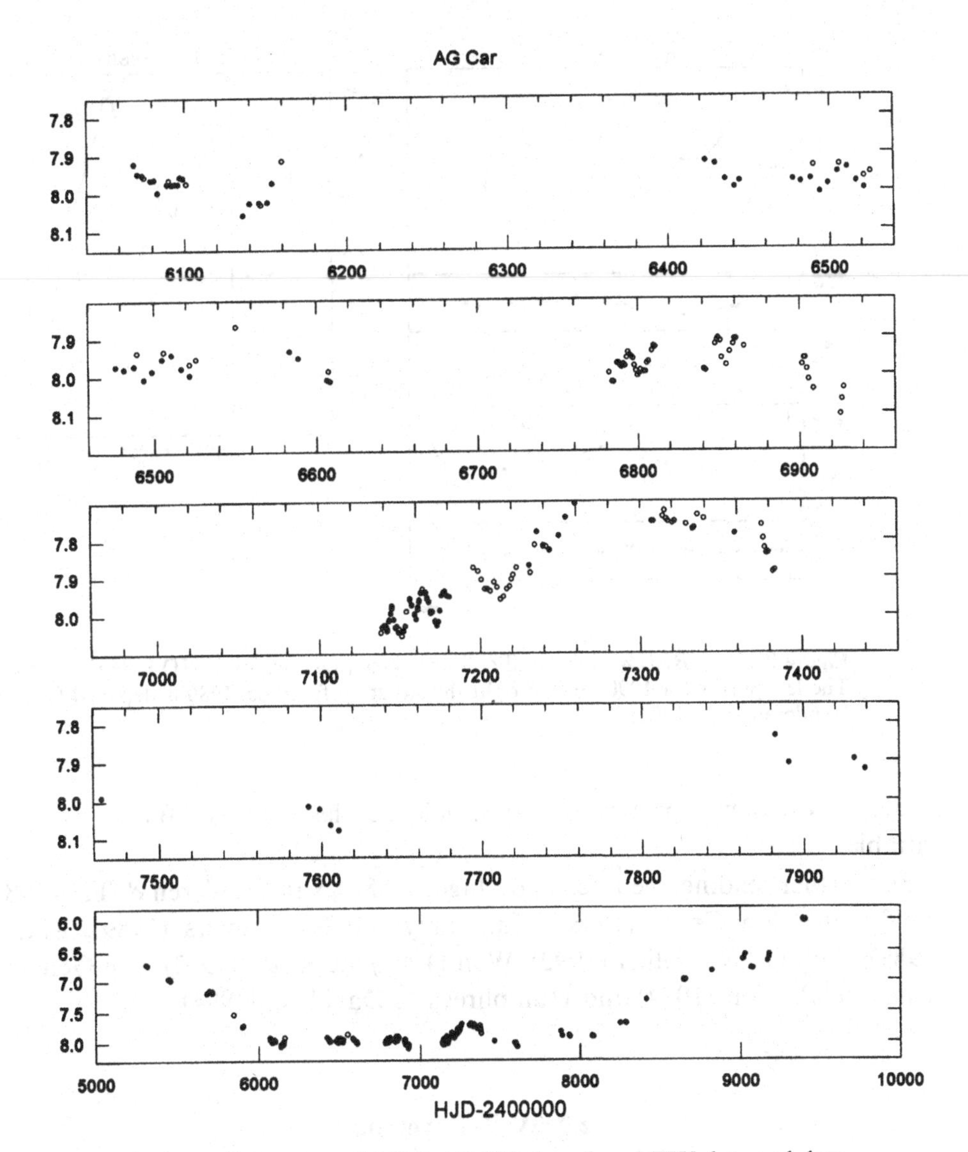

Figure 2.10 *V* light curve of AG Car (LBV) (based on LTPV data and data discussed by Spoon *et al.* 1994).

levels, in other words: permanent non-quiescence characterised by small [and large] S Dor eruptions which, in fact, are shell ejections

(iii) light curves which are relatively smooth compared to normal α Cyg variables

In contrast, HD 168625, a nearby very luminous supergiant, must be classified – on the basis of its known light history – as an α Cygni variable (see Fig. 3.2). The diagnostic based on the MLA–log $T_{\rm eff}$ diagram alone is not always sufficient

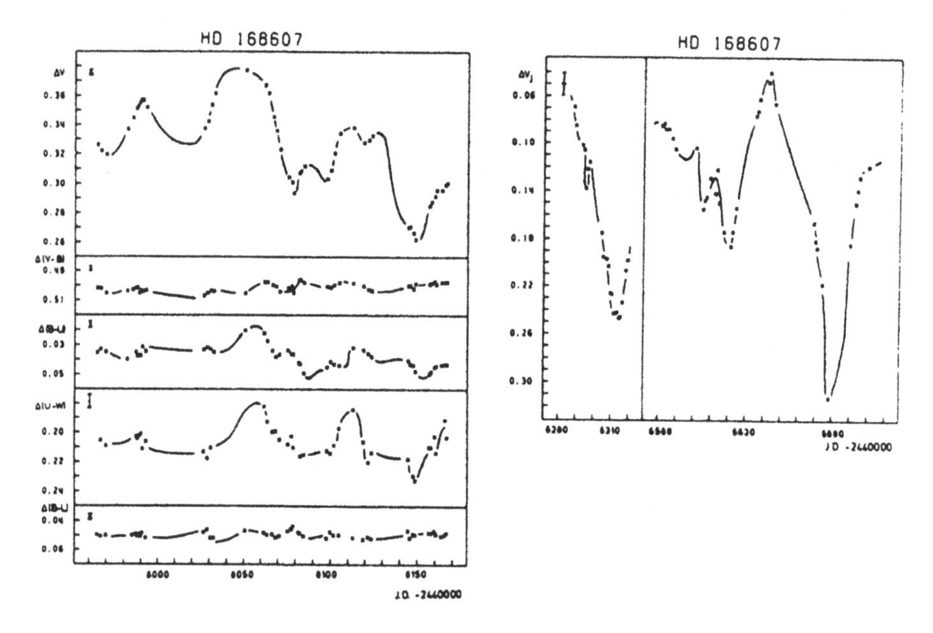

Figure 2.11 *VBLUW* light- and colour curves (log *I* scale) of HD 168607 (LBV). The left part is for 1990, on the right the situation for 1985–1986 is depicted (van Genderen *et al.* 1992).

to decide whether a particular star should be classified as LBV or as α Cyg variable.

For further reading, we refer to de Jager (1980), van Genderen & Thé (1984), Wolf (1986), de Groot (1988), Humphreys (1989), Lamers (1989), Maeder (1989), Wolf (1989), Hillier (1992), Wolf (1992), de Koter (1993), van Genderen *et al.* (1994), Wolf (1994) and Humphreys & Davidson (1994).

2.2 Wolf–Rayet stars

C. Sterken

Wolf–Rayet stars are very luminous hot Population I stars[1] of effective temperatures between 30 000 and 50 000 K with very peculiar spectra that are characterised by the presence of broad and strong C, N, O, He and Si emission lines along with absorption lines corresponding to spectral type O or B. WR stars are also notorious for their high mass-loss rate ($10^{-5} M_{\odot} yr^{-1}$). WR stars form an important evolutionary phase through which all massive stars above

1. There is some confusion in nomenclature, since the name is also used for central stars of planetary nebulae (emission Population II stars).

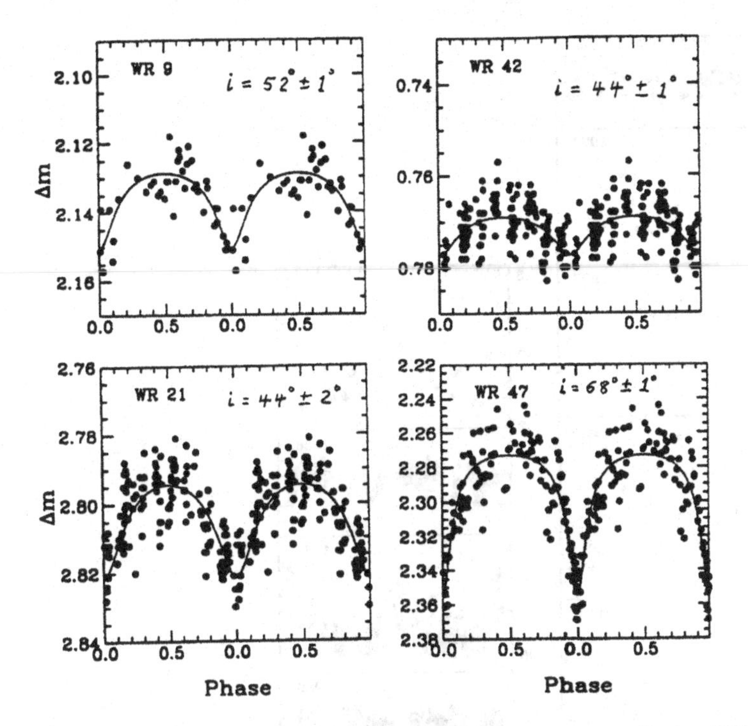

Figure 2.12 Phased visual light curves of four non-core eclipsing WR + O binaries; the fitted curves are from models of atmospheric transparency variations due to electron scattering of WR wind material along the line-of-sight of the O star (Moffat 1994) .

a certain limiting mass pass when going from the main sequence to the end of their lives.

The first objects of the group were discovered by Charles Wolf and Georges Rayet in 1866 (see Lührs 1991), during a visual spectroscopic survey in the same year that Secchi discovered, for the first time, emission lines in the spectrum of γ Cas, a Be star (see Section 3.3). Their discovery was presented at the Académie des Sciences by Le Verrier (Wolf & Rayet 1867). WR stars can be subdivided in two subclasses, the so-called carbon sequence (WC stars, WC4–WC9, with strong C and O emission lines), and the nitrogen sequence (WN stars, WN2–WN9, with strong He and N emission lines). The foundation of this classification is due to Payne (1933) and to Beals (1938), see also Smith (1968). The emission lines belong to a wide range of ionisation of highly-ionised elements. Many WR stars are double-lined spectroscopic binaries; only a few eclipsing binaries are known.

About half of the WR stars show light variations with amplitudes of several hundredths of a magnitude, but some WR stars are constant to within $0\overset{\rm m}{.}005$. Some light variations in WR binaries are caused by phase-dependent

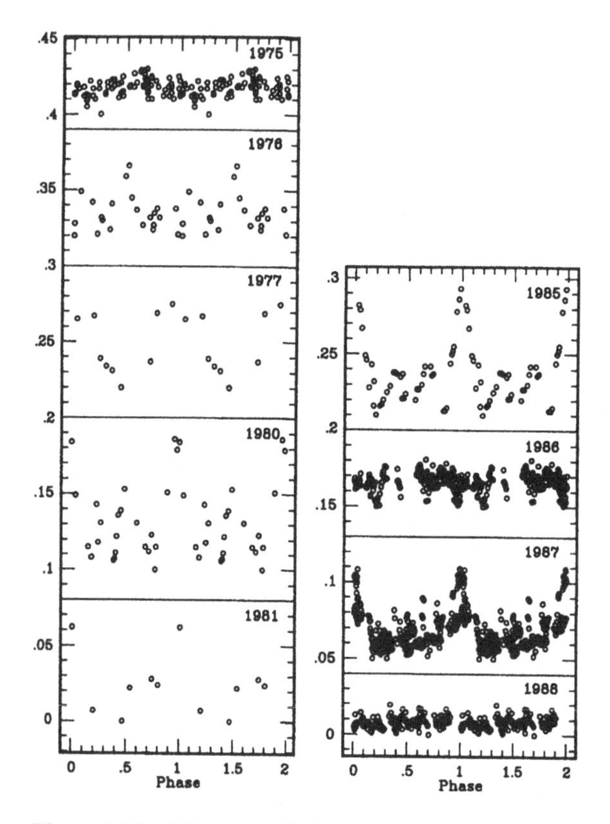

Figure 2.13 Montage of photometry of the WN5 star HD 50896 according to ephemeris 2440000.300 + 3.7658E (Balona *et al.* 1989).

occultation and tidal effects, with additional variability of stochastic nature (see, for example, Fig. 2.12, 2.13 and 2.14). Time scales of variation range from milliseconds to seconds (pulsars in a WR binary), from minutes to hours (flares and pulsations) and from hours to several days (their strongest variations) and years, with variable amplitudes; multiple periods are often present. Light curves of a complete sample of northern WR stars have been presented by Moffat & Shara (1986). For a discussion of observed light variability of WR stars, we refer to Vreux (1987). Light variations are also due to the variable strength of the emission lines, and the impact of such variations on the light transmitted by almost any existing filter system (broad-band as well as narrow-band photometric systems) is very difficult to separate from the variations of the underlying continuum – unless one uses specially-designed narrow-band filters that correspond to the emission-free parts of the continuum.[1] Moffat *et al.* (1989) consider WR stars (particularly the most luminous ones, of

1. Such a problem also occurs with LBV, Be star and symbiotic star photometry.

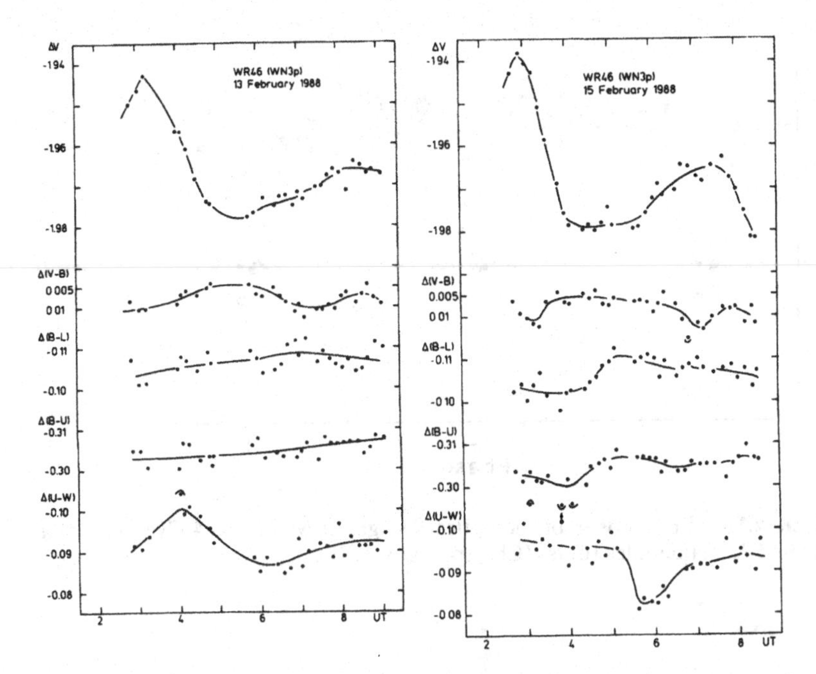

Figure 2.14 $V, V-B, B-L, B-U, U-W$ light curve of WR 46 as a function of UT (van Genderen *et al.* 1990b). The spread around the hand-drawn curves is caused by observational scatter.

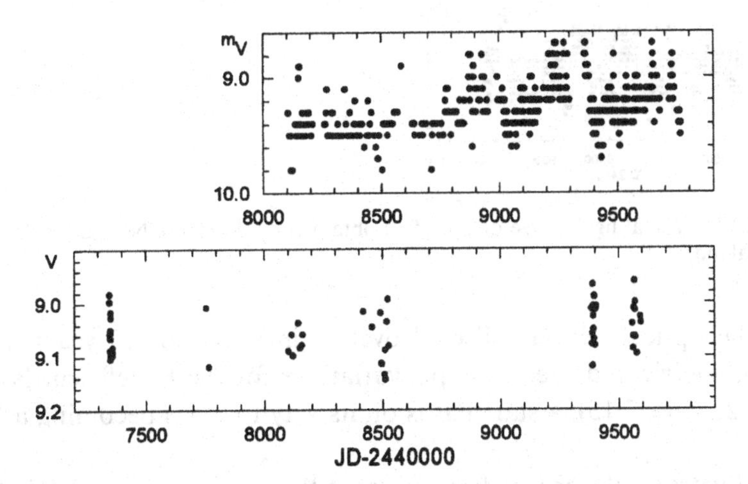

Figure 2.15 Photoelectric V light curve (bottom) and visual light curve (top) of HDE 326823; the visual data were obtained by A. Jones, the V data are LTPV data, see Sterken *et al.* (1995b).

subtypes WN6–9) as post-LBV stars (see also Section 2.1), indeed some WR stars (notably WN8–9) display a degree of variability that resembles that seen in LBVs. An illustration of such variations was recently given by Sterken et al. (1995b), who reported (on the basis of *uvby*, visual photometry, and spec-

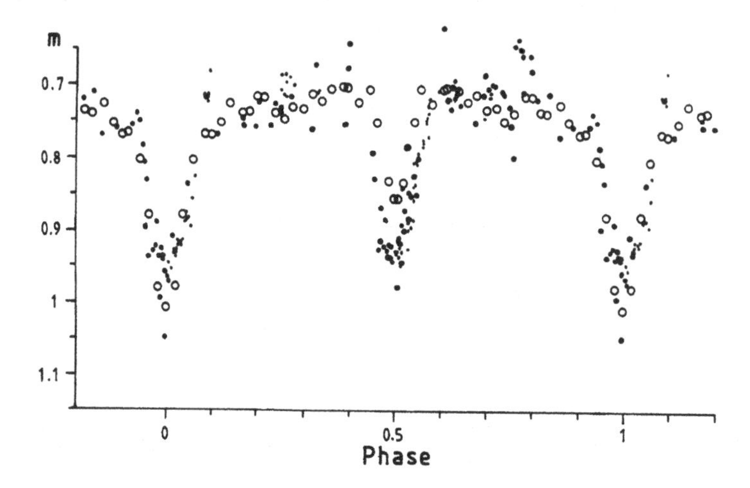

Figure 2.16 Comparison of the optical U-light curve of V 444 Cyg (circles) and infrared K magnitudes (dots) (Cherepaschuk 1992).

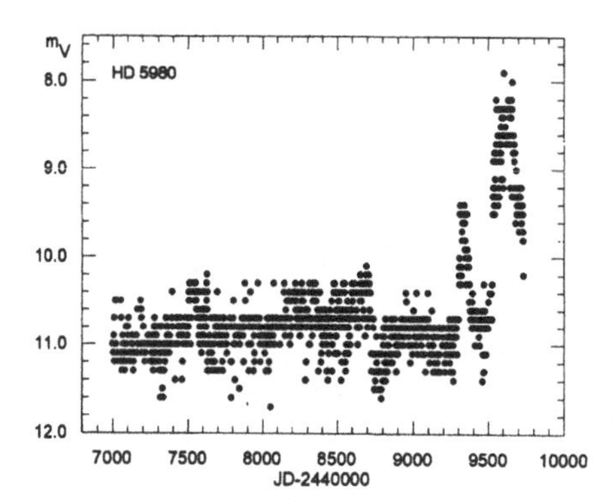

Figure 2.17 Visual light curve of HD 5980 obtained by A. Jones, Nelson, New Zealand.

trography in the optical region collected over a time span of 10 years) large amplitude photometric and spectroscopic variations for the H-deficient N-rich star HDE 326823 (Fig. 2.15), a star that is on its way towards becoming a WN star.

Figure 2.16 illustrates the eclipse light curve of the very interesting WN5+O6 WR eclipsing binary V 444 Cyg ($P = 4^{d}.2$), a key star for determining the structure of the extended atmosphere of a WR star (Cherepaschuk *et al.* 1984); width and depth of the eclipse of the WR star by the O6 star are increasing strongly with wavelength.

A most fascinating flare event was observed in HD 5980, the brightest WR eclipsing-binary pair in the SMC (spectral type OB? + WN3; the system has a strongly-eccentric orbit with $e = 0.47$, Breysacher & Perrier 1991). The

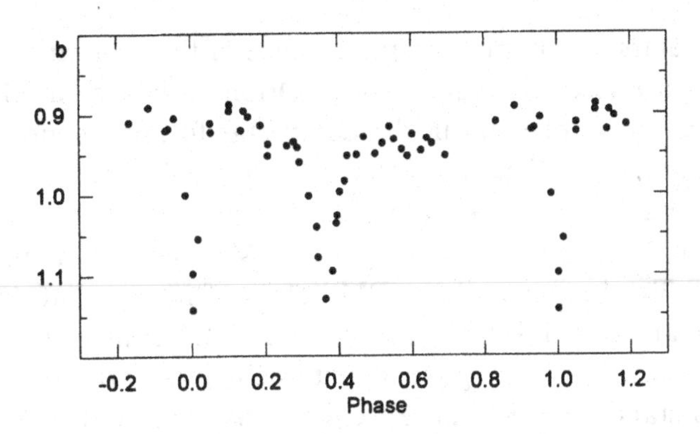

Figure 2.18 Strömgren differential b phase diagram of HD 5980 ($P = 19\overset{d}{.}266$, data from Breysacher & Perrier 1980). Note the unequal depth of primary and secondary minima, and the displacement of the secondary minimum, indicating strong eccentricity.

star brightened by one magnitude in July 1994 (Bateson 1994) and completely changed its spectrum to one similar to the spectrum of an LBV, with strong P-Cygni lines of He I and H (Barba & Niemela 1994) – see Figs. 2.17 and 2.18.

Recent reviews on WR stars and WR variability are van der Hucht (1992), Gosset *et al.* (1994) Maeder & Conti (1994), van der Hucht *et al.* (1994), and Moffat (1994).

2.3 Pre-main sequence stars

J. Krautter

Pre-main sequence (PMS) stars are newly-formed stars which have been born recently out of interstellar matter, and which have not yet reached a sufficiently high central temperature to ignite core hydrogen-burning, i.e. the central nuclear reactions that produce the energy in a 'normal' star. During their formation stage the PMS stars gain their energy by release of gravitational energy due to the so-called Kelvin–Helmholtz contraction.

PMS stars are usually classified as eruptive variables; in the *GCVS* they are subdivided in many subgroups (FU, IN, INA, INA, INT, INYY, and several others). However, this classification, which is based on morphological photometric properties only, is very inhomogenous and of only very limited value for a real physical classification, which should be based on the intrinsic physical properties of these objects. For instance, the RW Aur type variables (Is in the *GCVS*) are physically nothing other than low-mass PMS stars which are usually called T Tauri stars (InT in *GCVS*), or the FU Ori type variables (FU in

GCVS) are also T Tauri stars which are in a special stage of their evolution. All these kinds of variables are also sometimes called 'Orion variables' or 'Orion population' (one finds all types of them in the Orion star forming region), or 'nebular variables', since many of them are connected to nebulosity.

Now it is common practice to make a physical distinction between the PMS stars according to their mass: the low-mass PMS stars with masses $M \leq 3M_{\odot}$ are called T Tauri stars, and the PMS stars of intermediate mass ($4M_{\odot} \leq M \leq 8M_{\odot}$) are called Herbig Ae/Be stars. Of course, there is no clear-cut border between the two groups, but there is a rather smooth transition. The T Tauri stars are much more frequent than the Herbig Ae/Be stars: whereas we know more than a thousand T Tauri stars only some fifty Herbig Ae/Be stars are known.

T Tauri stars were first observed in the Taurus-Auriga dark cloud complex by Joy (1942) more than fifty years ago. He named this group of stars 'with late spectral types and emission lines resembling the solar chromosphere' after the brightest member, T Tau. The physical nature of the T Tauri stars was first recognized by V. Ambartsumian in 1947, who proposed 'that these objects are very young stars of low mass which have not yet reached the main sequence'. Later on, in 1962, another pioneer of the PMS research, G. Herbig, defined spectroscopic criteria for the T Tauri stars. Since then the opening of new spectral ranges (millimeter, infrared, ultraviolet, and X-ray) has given a wealth of new information, and has significantly changed our understanding of the T Tauri stars.

T Tauri stars are usually found in or near dark cloud complexes. They show a late type (G–M) photospheric absorption spectrum on which a continuous spectrum is superimposed. This continuous emission is sometimes so strong, that it is totally 'veiling' the photospheric absorption spectrum. Very characteristic for T Tauri stars are the emission lines which are, as immediately recognized by Joy (1942), similar to those of the solar chromosphere. The strongest lines are usually the hydrogen Balmer lines, other emission lines found are those of singly ionized metals like Ca II or Fe II, and lines of neutral helium (ionized helium is found in exceptional cases only). Some of the T Tauri stars also show forbidden emission lines which immediately tell us that these T Tauri stars must be associated with thin circumstellar matter (otherwise the forbidden lines could not originate). In part, the emission lines exhibit very complex profiles indicating mass flows in a circumstellar surrounding. Indicative of large amounts of circumstellar dust is strong excess radiation found at infrared and sub-millimeter wavelengths. A very important clue to the youth of the T Tauri stars is the strong absorption line of lithium at 6707 Å, a line not found in older stars since Li is depleted very rapidly in stellar atmospheres.

Whereas the T Tauri stars as described above are usually found on the basis of their strong emission lines and/or because of their strong infrared emission, X-ray observations with imaging X-ray telescopes have revealed a totally new class of T Tauri stars which have only weak emission lines and no or only very weak infrared excess emission. Hence, these stars – which were discovered on the basis of their strong coronal X-ray emission – were named 'weak-line' T Tauri stars as opposed to the 'classical' T Tauri stars. Apparently the 'weak-line' T Tauri stars have already lost most of their circumstellar material, but they are still gaining their energy from gravitational contraction and not from nuclear reactions in the stellar interior. Like the classical T Tauri stars, weak-line T Tauri stars show the strong Li absorption feature; actually, the optical identification is mainly based on the detection of this feature. Extended studies, especially based on data from the German X-ray satellite ROSAT, have shown that weak-line T Tauri stars are much more numerous than classical T Tauri stars. As the evolution of the PMS star continues, it moves towards the main sequence in the Hertzsprung-Russell diagram, still gaining energy by the gravitational contraction. If the temperature in the core is high enough to ignite hydrogen burning, the PMS star becomes a main sequence star, and the star formation process has ended.

The general picture has emerged that during the 'classical' T Tauri stage these objects are surrounded by circumstellar accretion discs which are supposed to be the places where later on planets could be formed; henceforth, they are sometimes also called 'protoplanetary' discs. The accretion discs are created during the star formation process in the interstellar clouds, and they are responsible for many of the characteristic properties of T Tauri stars. For instance, it is believed that outbursts of the FU Orionis type are due to instabilities in the accretion disc.

From the line profiles of the emission lines which very often show P Cygni type characteristics, one can conclude that T Tauri stars have stellar winds with velocities up to a few 100 $\mathrm{km\,s^{-1}}$. The mass loss rates are of the order of $10^{-8} - 10^{-7}\, M_{\odot}\,\mathrm{y}^{-1}$. Several T Tauri stars exhibit redward-displaced absorption components, the so-called 'inverse P Cygni profiles' (see also Section 2.1), which indicate infall of matter onto the stellar surface. The interaction of the T Tauri winds with the interstellar matter in which the T Tauri stars are embedded, is the physical reason for phenomena like Herbig-Haro objects, jets, and bipolar molecular outflows which are connected with the PMS stars.

Some of the very characteristic properties of T Tauri stars are the large variety of photometric variations and their very complex photometric behaviour in the sense that variations can have different character at different wavelength bands. (Indeed Joy first became interested in T Tauri stars in the early 1940s because of these photometric variations.) In most cases the variations are irregular over all wavelength bands from ultraviolet to infrared, except in the very far infrared range. The time scales of this type of variations can be minutes to decades or even centuries. Brightness amplitudes Δm of the variations can be as large as 5 mag or, in exceptional cases like FU Ori or EXor type outbursts, even somewhat larger. It is, therefore, not surprising, that for most of the variations the physical causes are still not yet quite clear. Possible mechanisms are starspots, instabilities in the circumstellar accretion disc, magnetic flare activity, or obscuration by dust clouds which orbit around the star in Keplerian orbits in the outer part of the circumstellar discs.

At least five basic types of optical variations can be distinguished:

(i) Irregular variations with large amplitudes on a long-term time scale. These variations are somewhat connected to the spectral appearance of the star, since they occur most frequently in T Tauri stars with strong emission lines in their spectrum and/or inverse P Cygni profiles (YY Ori stars)

(ii) FU Ori type outbursts (FUors) which show a strong increase by up to 6 magnitudes within a few months, and a slow decline on time scales of years to decades (see above)

(iii) EX Lup type outbursts in which the brightness increases like in the FU Ori type outbursts, within a few months, by up to 5 magnitudes, but decreases on about the same time scale. After their prototype EX Lupi (see Fig. 2.19), these objects are also named 'EXors'. EXor outburst could be similar to the outburst of FUors, but on a lower level

(iv) Irregular variations with low or moderate amplitude ($\Delta m \leq 1 - 2\,\mathrm{mag}$) on time scales of minutes to hours. Some of these outbursts could be due to solar-type flare activity

(v) Quasi-periodic variations on time scales of 1–10 days which should represent the rotation period. It is assumed that the brightness variations are due to starspots which are the same as sunspots, however on a rather different level, i.e., the starspots in T Tauri stars should cover a much larger fraction of the stellar surface than the sunspots on the sun. Typical amplitudes for this type of variations are a few tenths of a magnitude

Most of the Herbig Ae/Be stars, which were first described by G. Herbig in 1960, have, in many respects, the same characteristics as T Tauri stars: the emission line spectra in the optical and UV spectral range are very similar, i.e., they also show strong hydrogen Balmer lines, lines of singly-ionized metals with, in part, sometimes strong P Cygni profiles indicating mass outflows from the system, and a large infrared and sub-millimeter excess indicative of the

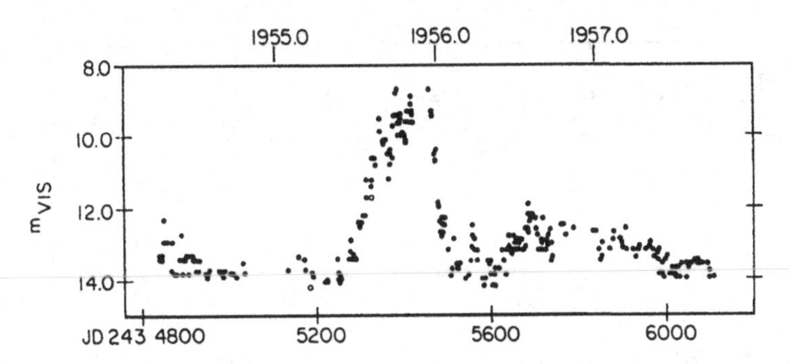

Figure 2.19 Visual magnitude estimates of EX Lupi, from Herbig (1977), see also Bateson & Jones (1957).

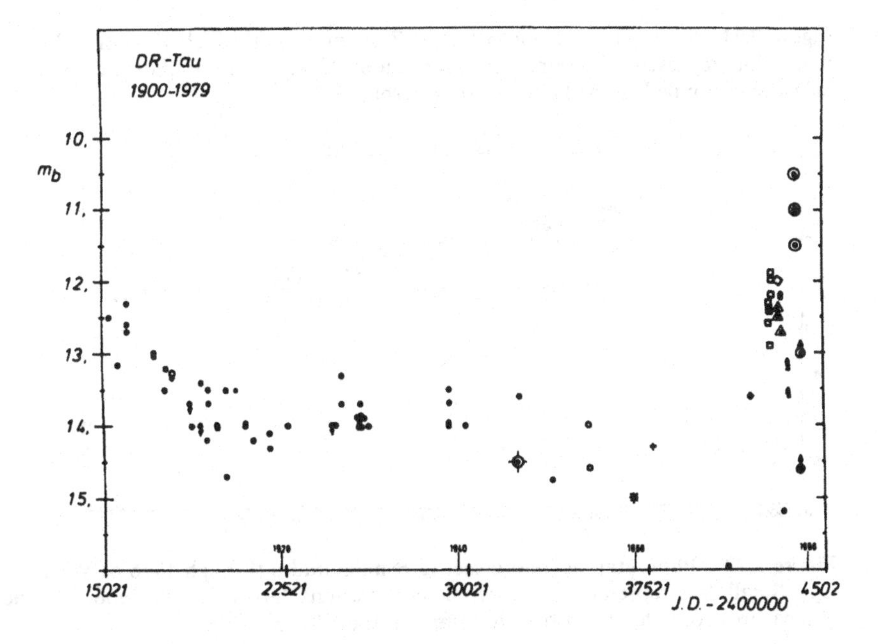

Figure 2.20 Light curve of DR Tau (T Tau), blue magnitudes taken from different observers and reduced to a common scale by Chavarria (1979).

presence of a substantial amount of mass of circumstellar dust. They are found at the same places in the sky as the T Tauri stars because of their physical association with dark cloud complexes. Herbig Ae/Be stars have – because of their higher mass – much higher luminosities than T Tauri stars.

One of the most remarkable properties of the Herbig Ae/Be stars is their irregular brightness variation which can be up to 3–4 magnitudes for HAeBe stars of spectral type later than A0. For the earlier objects the amplitudes of the variability are much lower, being up to a few tenths of a magnitude only. Again, the physical reasons for these variations, which occur on various time

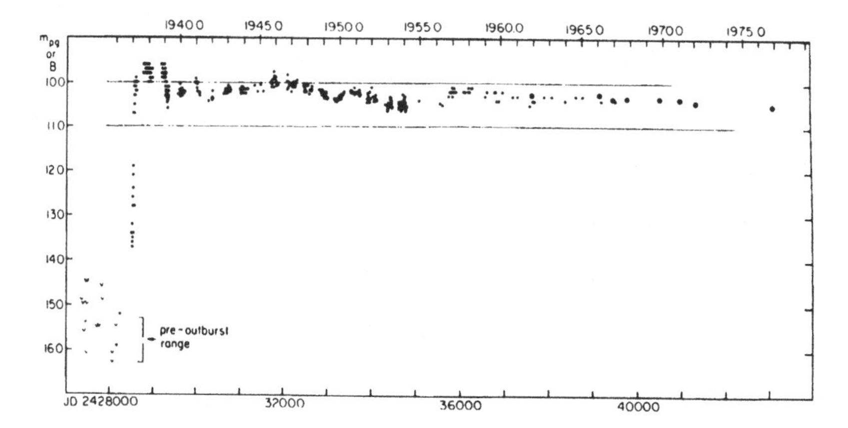

Figure 2.21 Photographic/photoelectric B magnitudes through 1976 of FU Ori. Small dots represent photographic observations, larger dots are photoelectric B measures compiled from the literature (Herbig 1977).

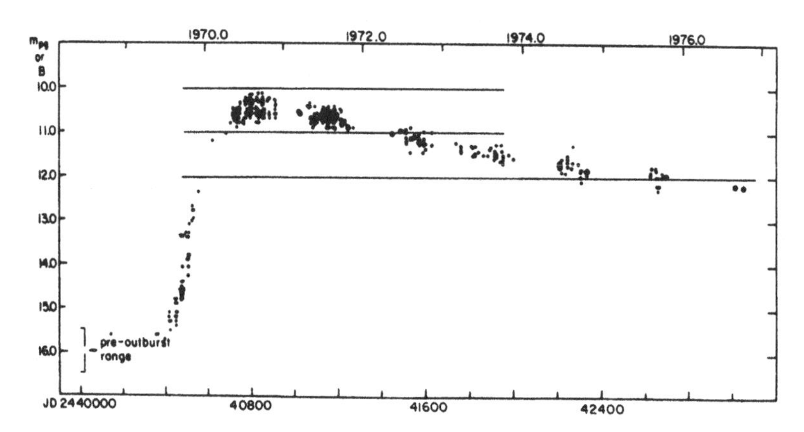

Figure 2.22 Photographic/photoelectric B magnitudes through 1976 of V 1057 Cyg. Small dots represent photographic observations, larger dots are photoelectric B measures compiled from the literature (Herbig 1977).

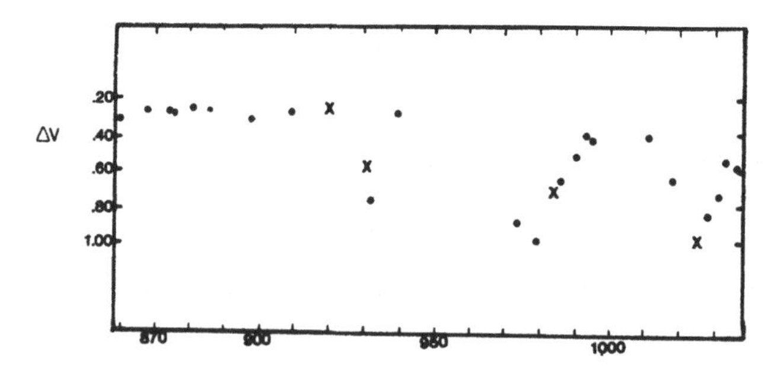

Figure 2.23 V light curve of SU Aur (T Tau), from Herbst *et al.* (1982). ΔV is the differential V magnitude with respect to BD +27 657 (crosses denote data of poorer quality). The X-axis is Julian date − 2 444 000.

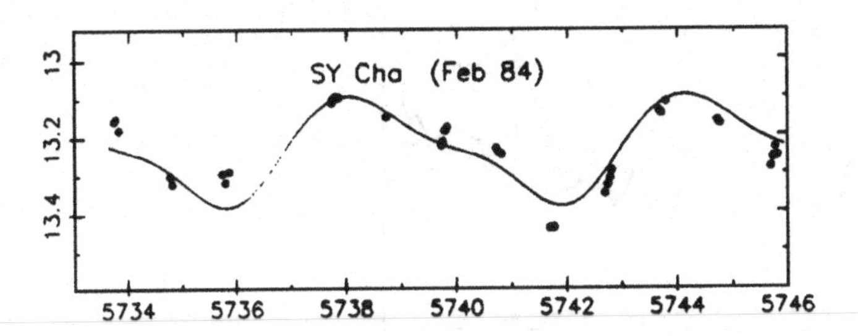

Figure 2.24 *V* light curve of SY Cha (T Tau), from Bouvier & Bertout (1989). The X-axis gives Julian days – 2 440 000.

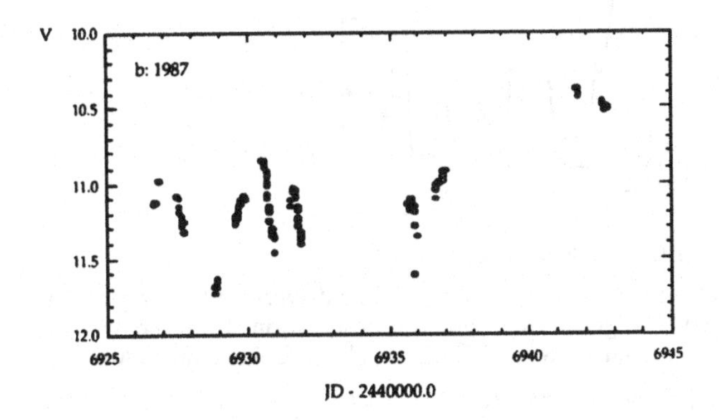

Figure 2.25 *V* light curve of RY Lup (T Tau), from Gahm *et al.* (1989).

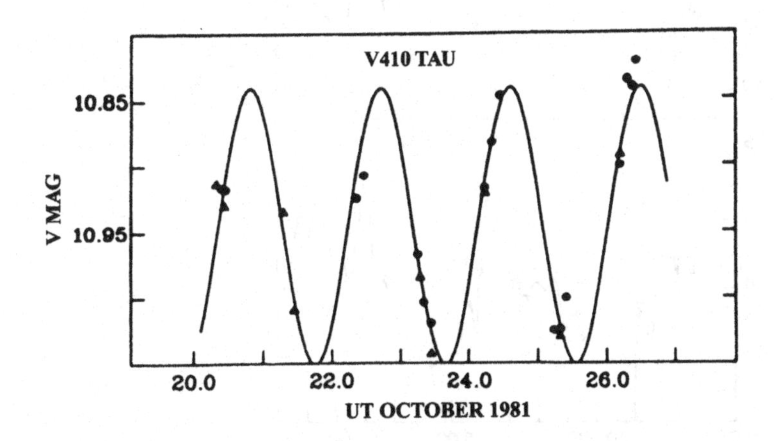

Figure 2.26 *V* light curve of V 410 Tau (T Tau) (Rydgren & Vrba 1983).

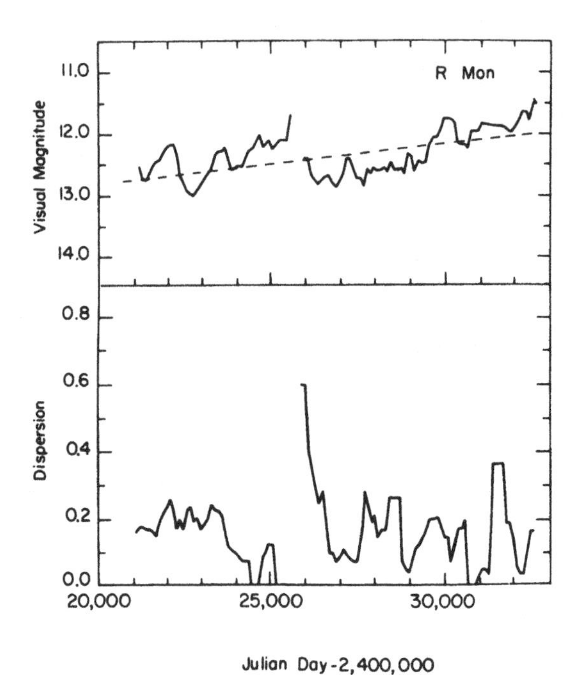

Figure 2.27 Visual magnitude (upper plot) and its dispersion (lower plot) of R Mon (HAeBe star). The graphs are based on observations from the *New Zealand Variable Star Circulars* from 1928 to 1955 (Bellingham & Rossano 1980).

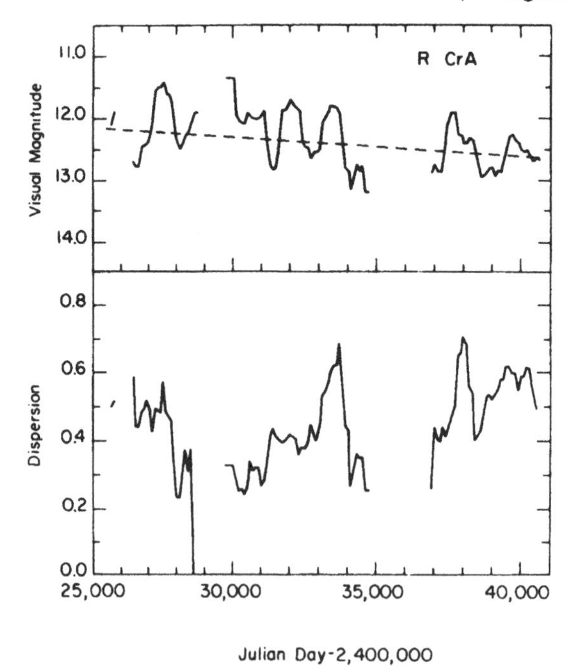

Figure 2.28 Visual magnitude (upper plot) and its dispersion (lower plot) of R CrA (HAeBe star). The graphs are based on observations from the *New Zealand Variable Star Circulars* from 1928 to 1955 (Bellingham & Rossano 1980).

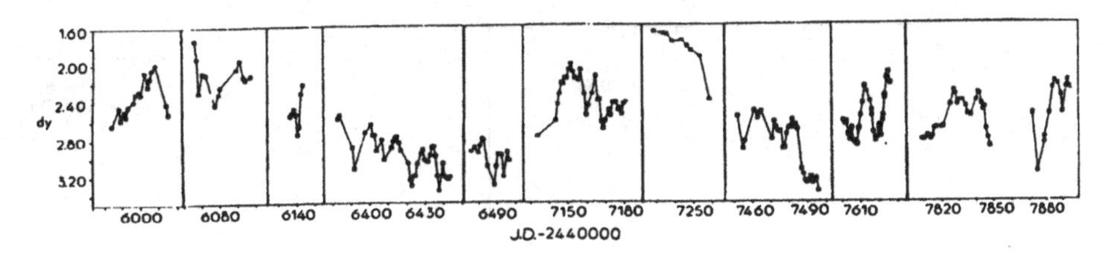

Figure 2.29 Strömgren $y = V$ magnitude variations of NX Pup (HAeBe star) (Bibo & Thé 1991).

scales are – as in the case of the T Tauri stars – rather unclear, but they should be in principle the same.

The sample of light curves is a selection representing the different types of photometric variations of the T Tauri and the Herbig Ae/Be stars. DR Tau (Fig. 2.20) is a very good example of irregular long-term variations with large amplitudes. The FUors are represented by FU Ori itself (Fig. 2.21), and by V 1057 Cyg (Fig. 2.22), the EXors again by the prototype EX Lup (Fig. 2.19). SU Aur (Fig. 2.23) shows low amplitude irregular variations, and the quasi-periodic variations are represented by SY Cha, RY Lup, and V 410 Tau (Figs. 2.24, 2.25 and 2.26). Of course, apart from the quasi-periodic variations these stars do show irregular variations too, which are, however, not shown here. As examples for Herbig Ae/Be stars average visual light curves of R Mon (Fig. 2.27) and R CrA (Fig. 2.28) are presented, along with individual measurement in the Strömgren y band of NX Pup (Fig. 2.29). The main characteristic of the Herbig Ae/Be variability is its irregularity!

For more general literature we refer to Appenzeller & Mundt (1989), Bertout (1989) and Reipurth (1989).

2.4 Flare stars

J. Krautter

Flare stars belong to the class of eruptive variables. They are late-type dwarf stars which undergo a sudden brightening at irregular time intervals. Their spectral type is K or M, but most are Me stars – that is, they show emission lines in their spectrum. The increase of the brightness can be more than 6 magnitudes. The amplitude of the flares increases with decreasing wavelength, i.e., it is stronger in the U band than in the V band. The time intervals between consecutive flares can be very different; they are usually between several hours and several days. The total energy in a flare can amount to 10^{34} erg ($= 10^{27}$ J). These flares are, in principle, the same kind of phenomenon as solar flares, but

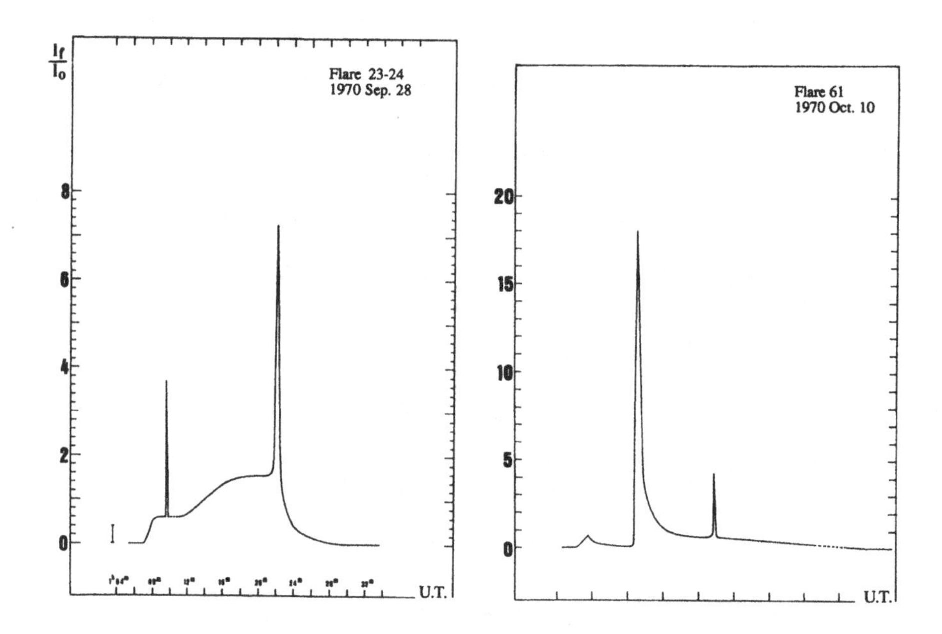

Figure 2.30 Relative intensity I_f/I_0 of B flare with respect to intensity of quiet star in the case of UV Cet (Cristaldo & Rodonó 1973, flares 23, 24 and 61).

with much higher energies involved – even the most energetic solar flares do not exceed 10^{31} erg. During a flare the stellar spectrum changes significantly, since emission lines appear that, in the quiescent state, are either not present at all or, in the case of the M emission line stars, are at least much weaker. The strongest emission lines are those of the hydrogen Balmer series, of helium and of singly-ionized metals like Fe II. The flare spectrum is very similar to the flare spectra of T Tauri stars. It is now clear that the flare phenomenon is directly connected to stellar solar-type activity. A similar kind of flare activity has been found in other variables like T Tauri stars, RS CVn stars, or in Algol type binaries.

Flare stars have been found in the solar neighbourhood as members of the galactic field; these field stars are also called UV Ceti stars after the prototype of this group. Very important has been the study of flare stars in open clusters and in associations; this group shows, on the average, a somewhat higher level of activity than the UV Ceti stars in the sense that the flares have a higher luminosity and do show up more often. It is clear, however, that the phenomenon is exactly the same as the ones seen in the UV Ceti stars; the different behaviour is due to age differences: flare stars in clusters and associations are, on the average, younger than UV Ceti stars. It has been found that the flare activity depends on the age of a star, and it decreases with

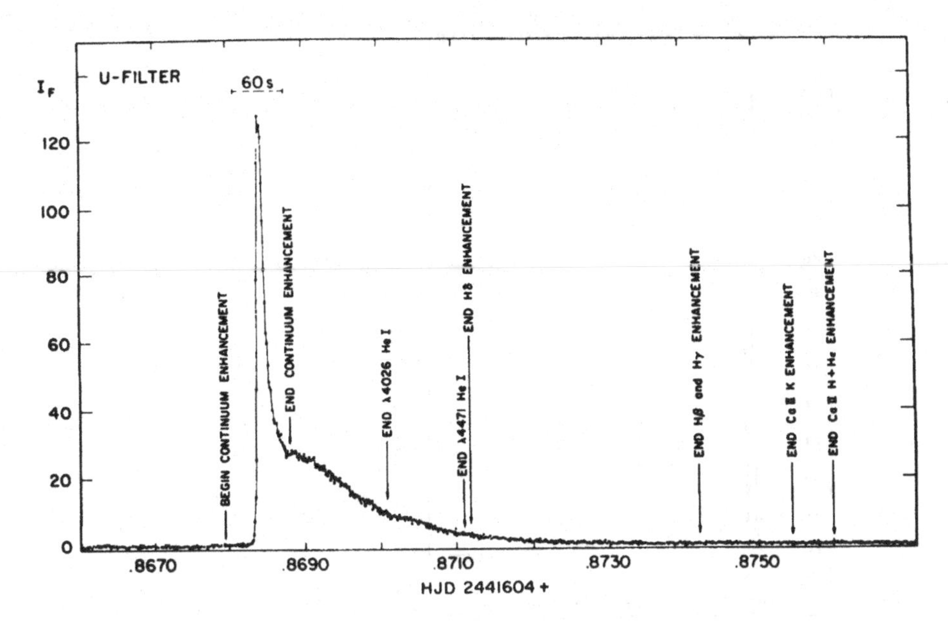

Figure 2.31 Flare intensity in *U* (per second) for UV Cet (Bopp & Moffett 1973). X-axis is time in fraction of Julian day.

increasing age. The investigations of the cluster flare stars have shown that essentially all late-type main-sequence stars in a young cluster are actually flare stars; one has to conclude that the flare stage is an evolutionary stage which all low-mass stars have to undergo for some time. The flare stage lasts from a few hundred million to about one billion years; its length depends somewhat on the mass of the star, the period being longer for stars of lower mass. The total number of flare stars in the Milky Way is very high; estimates range from one to ten billion. However, because of their low intrinsic luminosity, only nearby flare stars have been found.

The rise time of a flare, until the maximum brightness is reached, is very short; it is in the range from a few seconds to a few minutes. Gradients are usually in the range from 0.05–0.2 mag s^{-1}, but can in exceptional cases be as high as 2.8 mag s^{-1}, as was observed in a flare of UV Cet (Figs. 2.30 and 2.31). According to the decay time of a flare – that is, the time until the brightness returns to the pre-flare level – two different types can be distinguished, the 'long-decay' flares with decay times of one hour or more, and the 'impulsive' or 'spike' flares with decay times of a few minutes to a few tens of minutes. For both types of flares analogies with the sun exist. These morphological differences probably indicate real physical differences in the energy release process. As is evident from the light curves presented here (Figs. 2.32–2.36),

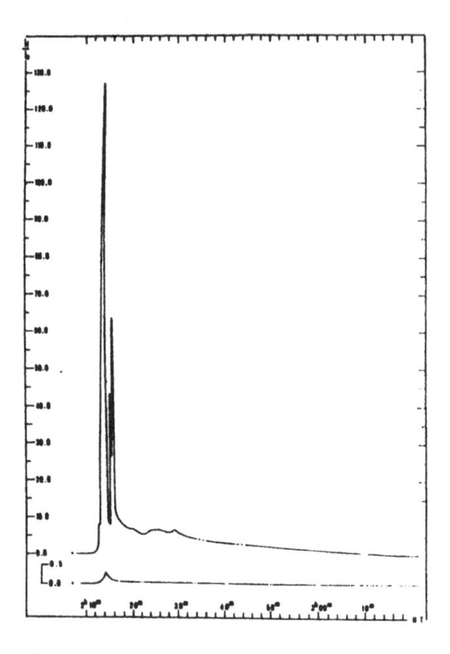

Figure 2.32 Relative intensity I_f/I_0 of B flare with respect to intensity of quiet star in the case of EV Lac (Cristaldo & Rodonó 1973, flare 30).

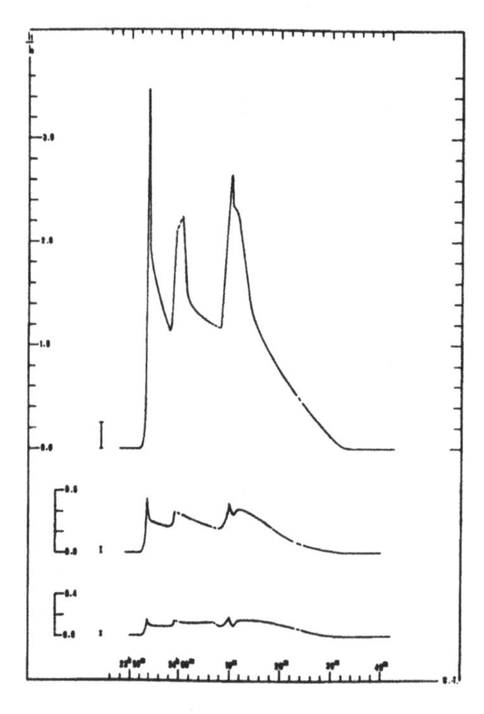

Figure 2.33 Relative intensity of flare with respect to intensity of quiet star in the case of EV Lac (Cristaldo & Rodonó 1973, flare 74) in U (top), B (middle), V (bottom).

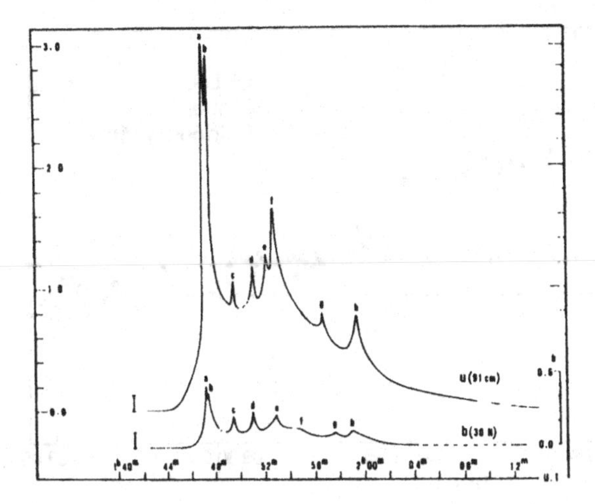

Figure 2.34 Relative intensity of *UB* flare with respect to intensity of quiet star in the case of EV Lac (Cristaldo & Rodonó 1973, flare 11).

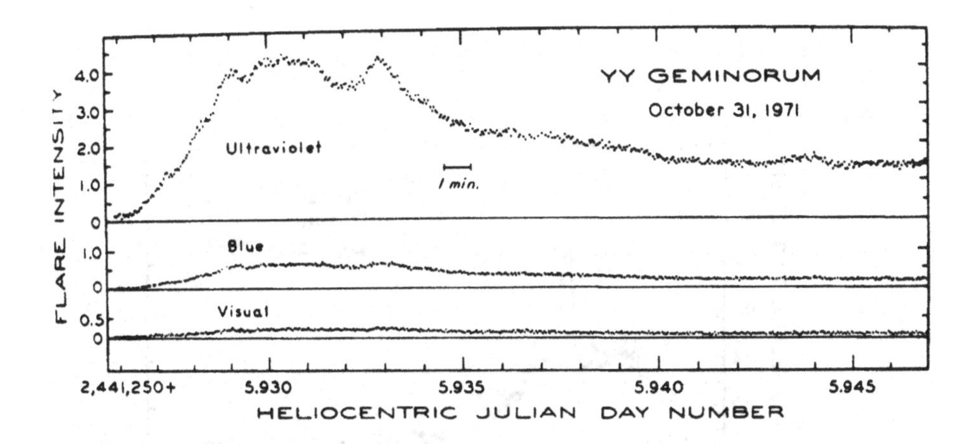

Figure 2.35 Flare intensity in *UBV* for YY Gem (Moffett 1974).

a wide variety of substructures have been observed in flare light curves, a diversity which, in most cases, is only very poorly understood.

Flares have been observed over wide spectral ranges from the radio to the X-ray regime. X-ray observations, especially, have turned out to be a very powerful tool to study flare eruptions, since flares are high-energy processes which emit a significant part of their energy in the X-ray regime. During X-ray observations, a totally new type of short-term flare activity has been found: the so called micro-flares, whose intensities are ten to a hundred times less than an ordinary flare. On the other hand, since micro-flares are a very frequent phenomenon, the total integrated energy released by this kind of flare is rather high. In the

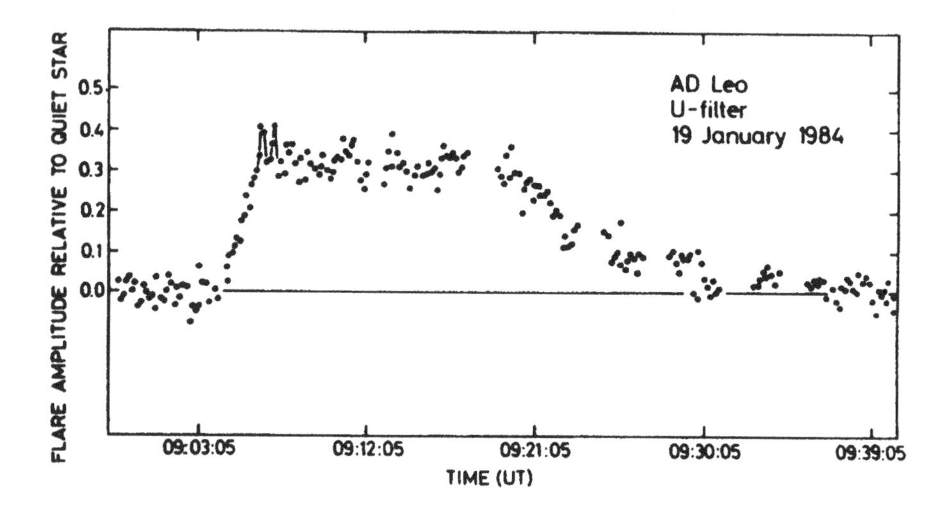

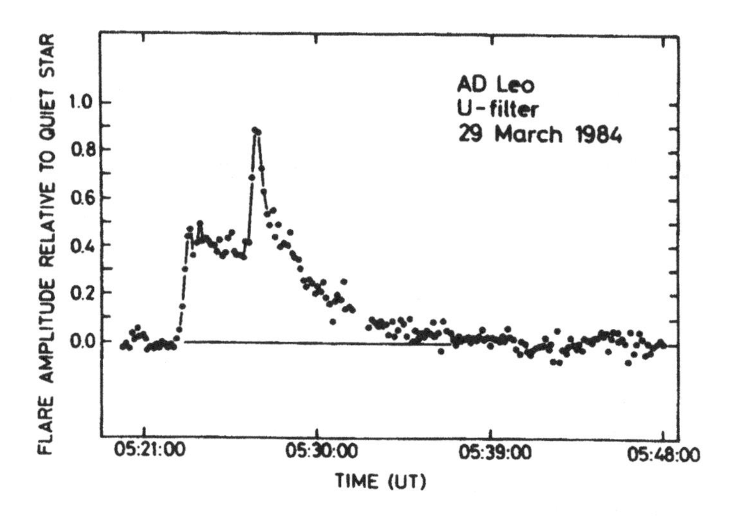

Figure 2.36 Flare amplitude relative to quiet star for AD Leo. The upper figure represents an unusal plateau-dominated flare, whereas the lower one represents a complex flare with an interrupted plateau phase (Petterson *et al.* 1986).

optical spectral range, micro-flares can not be observed yet. X-ray observations yielded the possibility to obtain estimates on temperature, electron density, and size of the flare emitting regions. Typically, the temperature is between 20 and 30 000 000 K, the electron density is in the range $10^{11} - 10^{12}\,\mathrm{cm}^{-3}$, and the size of the emitting region is about $10^{27} - 10^{28}\,\mathrm{cm}^{3}$.

At present the physical mechanism for flare activity is still very poorly understood. Any successful model for flare activity has to explain where the energy comes from, how this energy is stored in the stellar atmosphere, how this energy is released within a few seconds, and how the surrounding regions are heated to the high temperatures. It is now generally believed that flares are one manifestation of stellar activity, for which magnetic fields, as well as stellar rotation, play a crucial role. Most promising are those models in which the flare energy is released by a very rapid change in the configuration of the stellar magnetic field; generally this kind of physical process is called 'reconnection'.

The flare light curves presented here show the two different types of flares, 'long-decay' and 'spike' flares, as well as many different kinds of substructures. From the light curves of UV Ceti it is immediately obvious that a large variety of flare behaviour can be found in a single star.

For more general reading we refer to Mirzoyan *et al.* (1989) and to Haisch *et al.* (1991).

2.5 R Coronae Borealis variables

M.W. Feast

The R Corona Borealis (RCB) variables, named after the brightest star in the class, are a rare group of objects with perhaps only about 30 true members known. They spend most of their time close to maximum light but at (apparently) unpredictable intervals they undergo spectacular drops in brightness of up to nine magnitudes in the visible (Fig. 2.37). A typical major decline of this sort has a rapid phase (a drop of perhaps 4 mag in 25 days) followed by a more gradual decline (about 3 mag in 120 days). However, there is a great deal of variety in the form of the light curves from one decline to another. Much less spectacular drops in brightness can also take place. The recovery generally begins soon after the minimum is reached and is often much slower than the decline. It may take 1 to 3 years before the star reaches maximum again. New declines may occur before the recovery is complete or a new major decline may not take place for many years.

The stars are carbon rich and hydrogen poor and the decline in light is attributed to the formation of a cloud of carbon particles in ejecta from the stellar atmosphere. There is still considerable uncertainty as to whether the particles are formed close to the stellar photosphere or much further away (perhaps about 20 stellar radii above the surface of the star). Near-infrared observations have been important in understanding the RCB stars. They strongly suggest that the formation of the carbon particles during a major

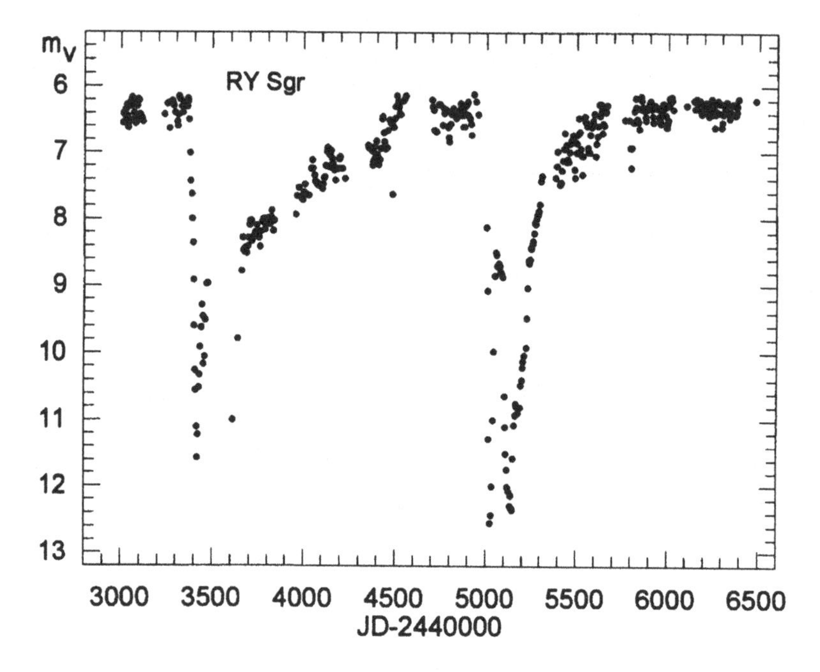

Figure 2.37 Visual light curve of RY Sgr (RCB). Data from AAVSO.

light-decline is taking place only close to the line of sight and not uniformly over the whole star. It then seems plausible that there will be times when carbon particle formation is taking place out of the line of sight and thus without influencing the optical brightness. In this model the star is ejecting puffs of soot (or other forms of carbon) in random directions and occasionally these are in the line of sight. The infrared observations show that RCB stars are surrounded by dust shells (presumably of carbon particles) at a temperature of about 800 K. The infrared luminosities of these shells vary with quasi-periods of 1000–2000 days.

Most RCB stars show variability at optical wavelengths of a few tenths of a magnitude when not undergoing an obscuration event. In general this variation seems to be semi-regular in nature and often has a quasi-period of the order of a month. RY Sgr (Fig. 2.38) is somewhat atypical in showing a rather regular variation with an amplitude of about 0.5 mag and a well-determined period of about 38 days. Radial velocity observations show that this star (at least) is pulsating and infrared observations show that these pulsations continue uninterrupted when the star goes into an optical minimum. It is generally assumed that all members of the class are pulsating and that the initial ejection of material from the star is connected with pulsational instability. This instability may include the propagation of shock waves through the stellar

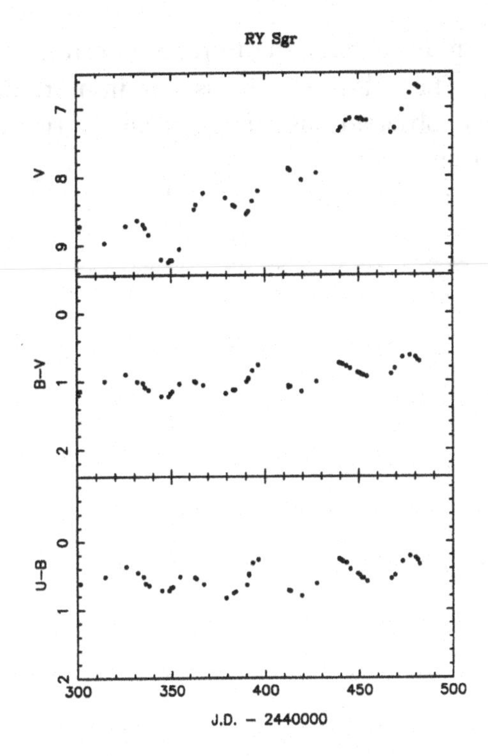

Figure 2.38 $V, B-V, U-B$ light curves of RY Sgr (RCB) during recovery from an obscuration event. The period is 38ᵈ.6 (Alexander *et al.* 1972).

atmosphere. There is some evidence that obscuration events begin at a specific phase of the pulsation cycle but this is not entirely conclusive and the details of the mass ejection process and particle formation remain obscure.

Three RCB stars in the Large Magellanic Cloud show that these stars are supergiants (absolute visual magnitudes of about −5). The best known RCB stars have surface temperatures similar to F, G or K stars with some objects possibly being cooler. There are also some rare hot RCB stars as well (spectral types B and A). The evolutionary state of the RCB stars and the relation of the hot ones to the cooler ones is not fully understood. One possibility is that they are in a rapid evolutionary phase from the top of the AGB towards the planetary nebula and white dwarf stage.

The figures illustrate the typical behaviour of RCB stars at visual wavelengths. Both figures are for RY Sgr which has probably been more extensively studied than any of the others. Figure 2.37 shows a light curve assembled by the AAVSO from visual observations of many amateur astronomers and extends over about 10 years. It shows two major declines and recoveries of the variable. Evidence of the 38 day pulsation cycle can also be seen. Much of the broad scatter in the points near maximum light is due to random errors in the eye-estimates of the star's brightness. The pulsation cycle is seen in more detail in Fig. 2.38 which shows photoelectric observations made during the rise back to maximum after the 1967 obscuration event.

Further useful references are Alexander *et al.* (1972), Feast (1975) and Hunger *et al.* (1986).

3
Pulsating variables

3.1 α Cygni variables

C. Sterken

According to the nomenclature of the *GCVS*, luminous variable B and A supergiants are called α Cygni variables, and are classified among the pulsating variables. The class also includes massive O and late type stars, since these belong to the same evolutionary sequence. In the MK spectral-classification system, they have luminosity classes Ib, Iab, Ia and Ia^+ (in increasing order of luminosity). The most luminous supergiants are also called 'hypergiants' – these are, in fact, Luminous Blue Variables (LBVs). Ia supergiants are pre-LBV objects, therefore we also refer to Section 2.1 for all details that are related to both groups of variables. All OBA supergiants are variable (Rosendhal & Snowden 1971, Maeder & Rufener 1972, Sterken 1977). The amplitudes of the most luminous supergiants resemble the microvariations observed in LBVs during quiescence, the level of variability increases towards higher luminosities for all spectral classes.

Pulsational instability accounts to some extent for the semi-regular variations (Leitherer *et al.* 1985, Wolf 1986) – it should also be noted that the β Cep instability strip widens into the supergiant region.[1] The amplitudes of the variations seem to increase with the time scales at which they occur.

HD 57060 = UW CMa and HD 167971 are two interesting cases of microvariations. HD 57060 (Fig. 3.1) is a binary consisting of an O8 supergiant star and an O or B type main-sequence star in synchronous revolution with a period of $4^{d}.39$. The system shows abnormally strong microvariations which may be connected with the strong distortion of the star filling its Roche lobe. HD 167971 (Fig. 3.1) is an interacting binary consisting of two main-sequence

1. For theoretical support for this viewpoint, see Dziembowski (1994); for observational evidence, see Sterken (1988), Sterken *et al.* (1996).

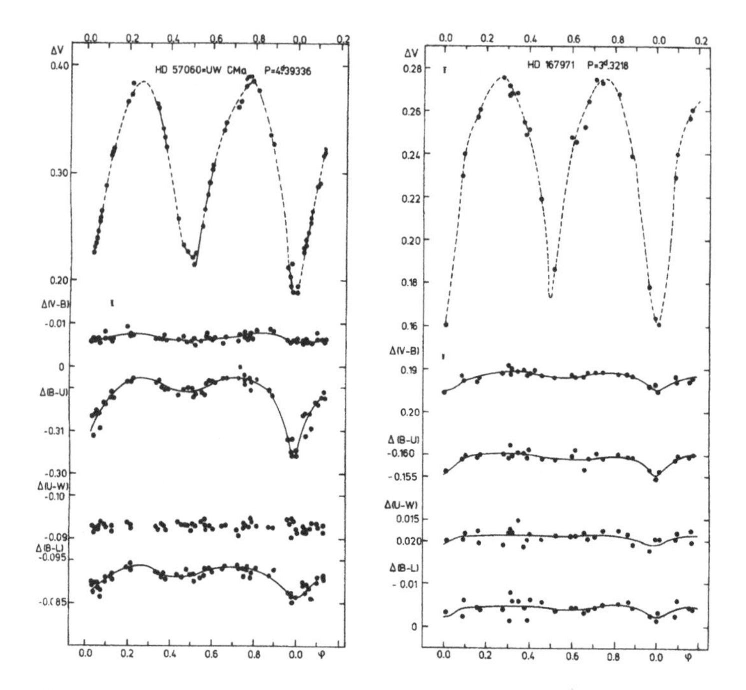

Figure 3.1 *VBLUW* light and colour phase diagrams of HD 57060 (ephemeris 2440877.563 + 4.39336*E*) and HD 167971 (ephemeris 2445554.97 + 3.33218*E*). The Y-axis is expressed in $\log I$ units with respect to the comparison star HD 58612 (for HD 57060) and HD 170719 (for HD 167971). The small vertical error bar in the middle indicates the mean error per data point for all colour curves (van Genderen *et al.* 1988).

O stars, with extra light from a third O8 supergiant which may or may not be a member of the system.

HD 168625 (Fig. 3.2) is a very luminous supergiant near the LBV HD 168607 (Section 2.1), but is still classified as an α Cyg variable, due to its low-amplitude variations. In fact, we most probably deal here with a case of a dormant LBV, as is possibly the case for HD 33579 (Fig. 3.3).

Note that some α Cyg variables display variability typical for other classes. One such example is Cyg X-1, an O9.7 Iab supergiant with a neutron star or black-hole secondary. The system is not only an X-ray source, it is also an ellipsoidal variable with a reflection effect due to X-ray heating of the surface of the optical star (Balog et al. 1981, see also Cherepashchuk *et al.* 1972).

ζ^1 Sco (HD 152236) is one example of an α Cyg variable that also can be considered an LBV, since it has all the characteristics as mentioned by Humphreys & Davidson (1994) of LBVs, but it has not been observed at high time resolution to be in an eruption 'à la η Carinae', and it is also not listed in Humphreys & Davidson's Table 2 of candidate LBVs. It is clear, though,

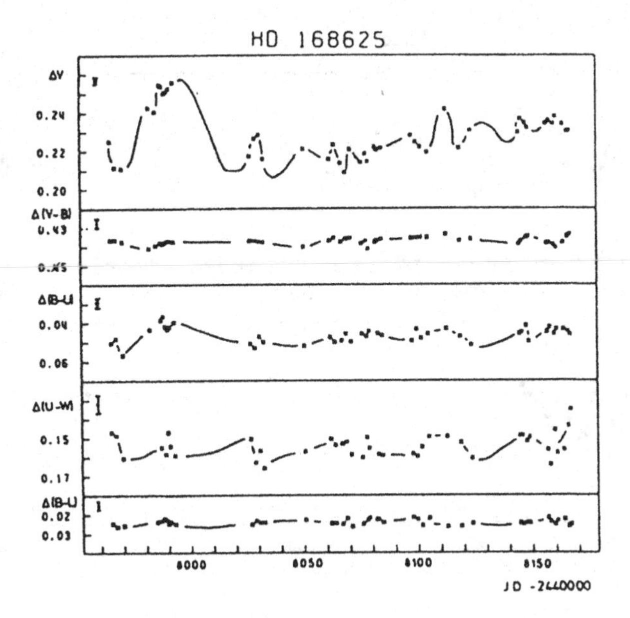

Figure 3.2 Differential *VBLUW* light- and colour curves (log *I* units) of HD 168625 for 1990 (van Genderen *et al.* 1992).

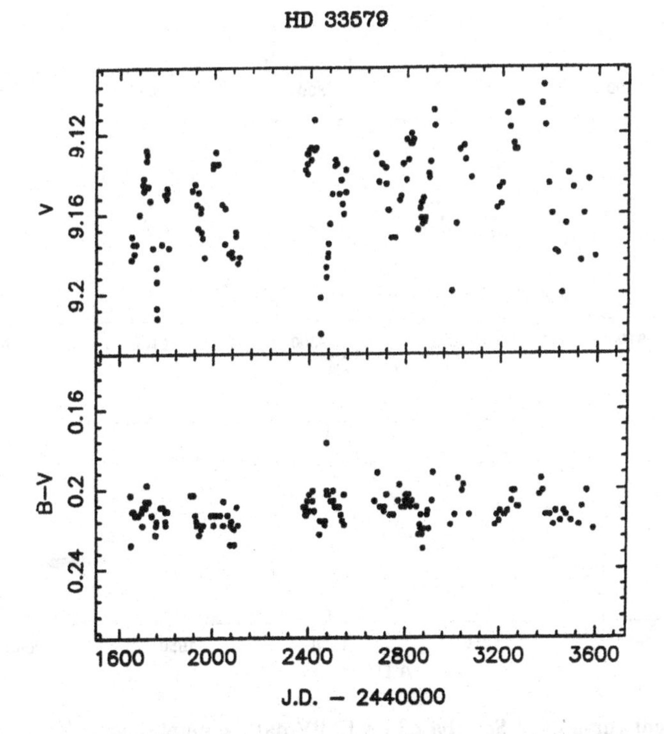

Figure 3.3 $V, B-V$ (derived by transformation from *VBLUW* data) light curve of HD 33579 (van Genderen 1979). Comparison star is HD 33486.

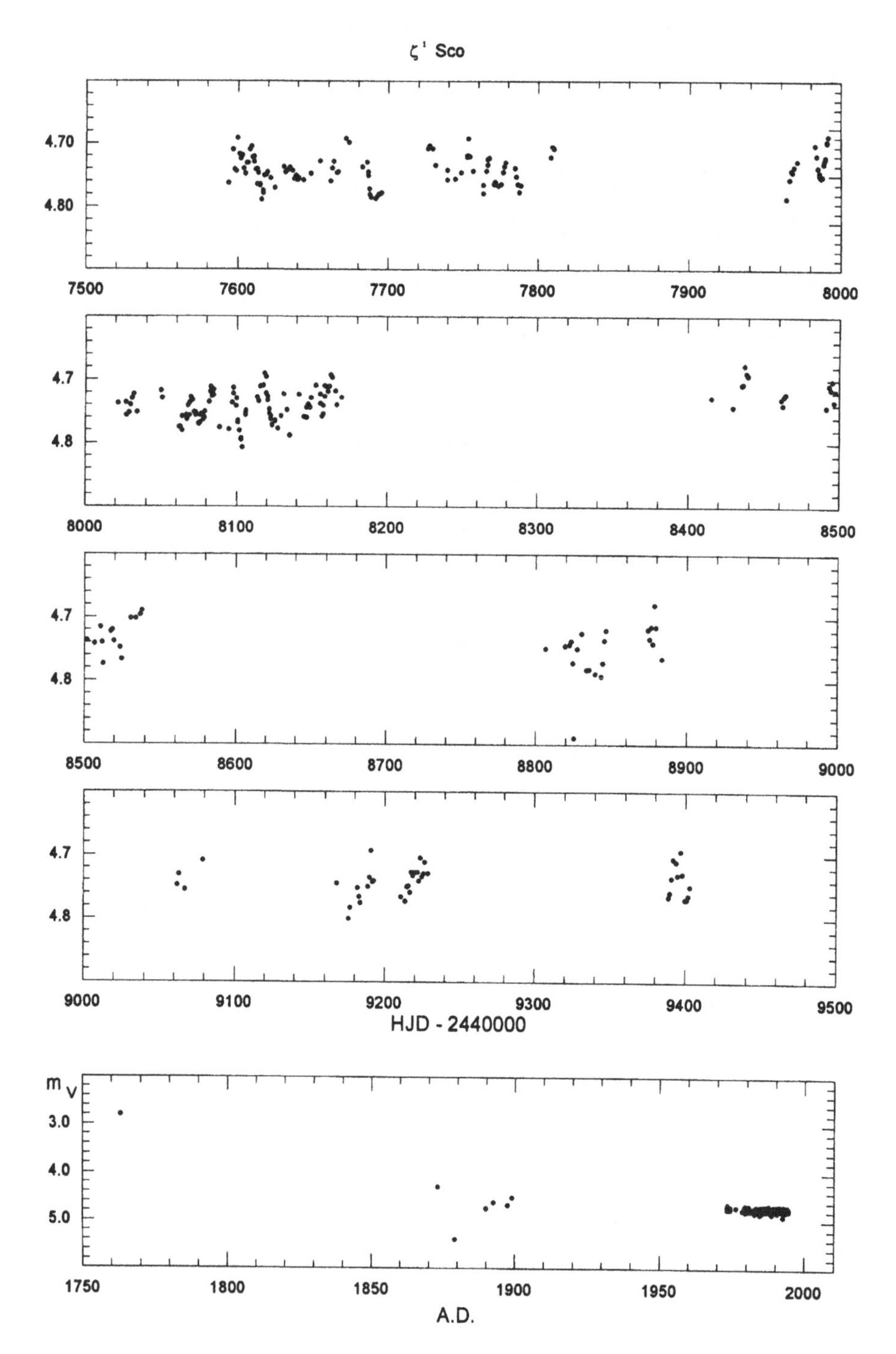

Figure 3.4 *V* light curve of ζ^1 Sco (based on LTPV data, unpublished measurements by Sterken 1977, Burki et al. 1982, van Genderen 1995 and several ancient catalogues).

that in the remote past, the star has been, at times, almost two full magnitudes brighter than we see it now (see Hertzog 1992 for other examples). Figure 3.4 shows the photometric microvariations, but also the historical light curve based on archived data (unpublished measurements by Sterken 1977, Burki et al. 1982, van Genderen 1995) and ancient catalogues (de La Caille 1763, Houzeau 1878, Gould 1879 and Zinner 1926).

For further reading, we refer to de Jager (1980), Lamers (1989) and Wolf (1986, 1989, 1994).

3.2 β Cephei variables

C. Sterken

The β Cephei variables are a group of apparently normal early B giants and subgiants which exhibit coherent short-period light and radial-velocity variations. The periods (two to seven hours) are too short to be explained by purely geometric effects (such as rotation and/or binary motion), and it was soon recognised that the only remaining explanation was one in terms of stellar pulsation. The interest of these variables for theoretical astrophysics, for a long time, was in the fact that theorists were not able to find a consistent explanation for the pulsational behaviour of these stars, so that the problem of the unknown driving mechanism for β Cep star pulsation remained one of the outstanding problems of stellar pulsation theory. But this situation has now changed.

The variability of the radial velocity of β Cep was discovered by Frost at the beginning of this century at Yerkes Observatory; he also determined that the

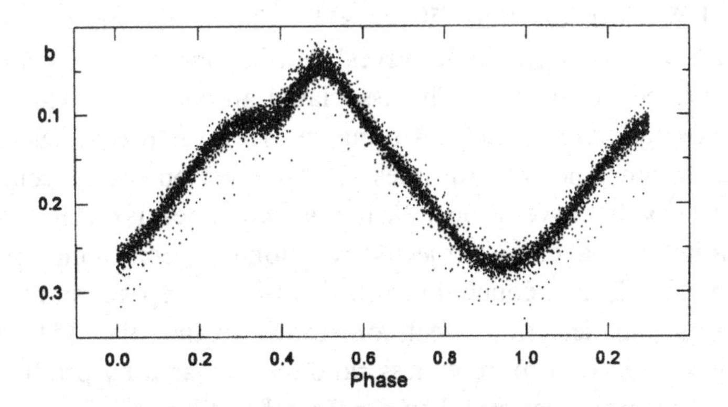

Figure 3.5 Differential *b* phase diagram of BW Vul, $P = 0^{d}.2010425$ (based on more than 6000 measurements obtained in 1982 from a coordinated campaign involving 15 sites, Sterken *et al.* 1986). The comparison star is HD 198820.

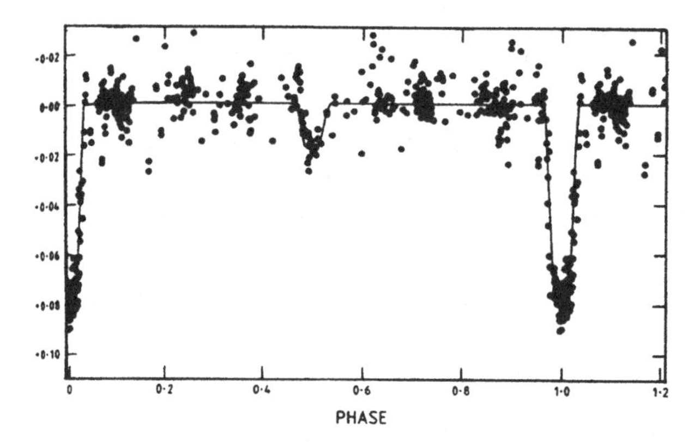

Figure 3.6 Differential *B* phase diagram of HD 92024 = NGC 3293–5 with orbital phase, $P = 8\overset{d}{.}323$ (Engelbrecht & Balona 1986). The comparison star is NGC 3293–20.

period of variability was $4^h 34^m$ (Frost 1902, 1906). Six years later, Guthnick revealed that with the velocity variation there was associated a light variation with the same period and with an amplitude of $0\overset{m}{.}05$ (Guthnick 1913). In 1908 Albrecht at Lick Observatory (Albrecht 1908) found that β CMa showed similar variations, and it turned out that this star became the first well-studied member of the group of β Cep variables. For several decades these stars were labeled 'β Canis Majoris stars', but there is now a very general tendency to use the historically correct name for this class of stars. The actual number of known β Cep stars exceeds 50, for a detailed review, see Sterken & Jerzykiewicz (1994).

The range of light variations in β Cep stars is – except for BW Vul – less than $0\overset{m}{.}1$ in the visible (see Fig. 3.5), and the radial velocity range is (except for σ Sco and for BW Vul) less than 50 $km\,s^{-1}$. The variations are often monoperiodic (with quasi-sinusoidal light curves), but in some cases harmonics of the fundamental period are present. The associated amplitudes of variation for these secondary periods are smaller. Among the 15 investigated cases, 13 show line profile variations: the spectral lines are broadest on the descending branch of the radial velocity curve. Line splitting is sometimes associated with these profile variations. The average projected rotational velocity amounts to 66 km sec^{-1}, but one should not conclude from this that β Cep stars are slow rotators, since a very strong selection effect contributes to this bias. (The first β Cep stars were discovered in a more or less random way as a by-product of stellar spectroscopy – on spectra often taken for the sake of kinematical studies – and since the amplitudes of these radial-velocity variations are relatively small, such variations could only be detected in sharp-lined stars.) Later discoveries

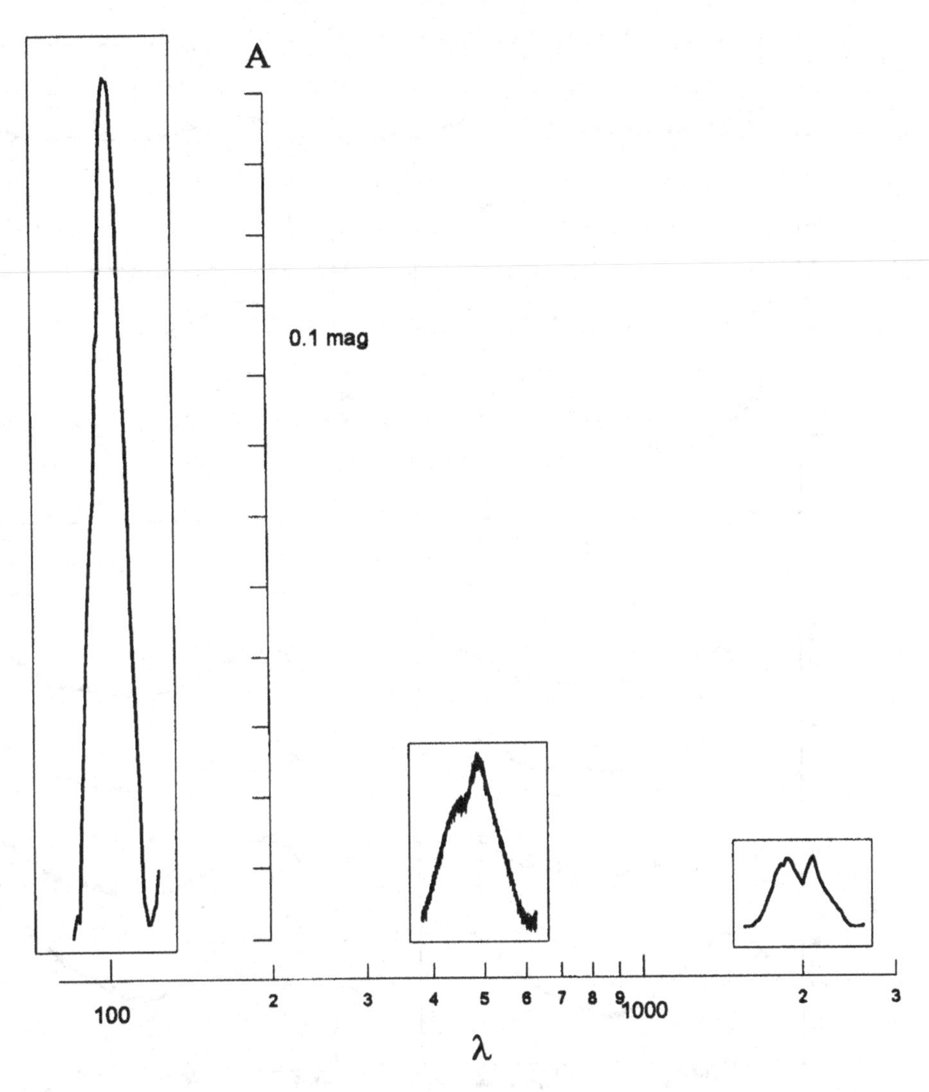

Figure 3.7 A composition of ultraviolet, visual and near-infrared phase diagrams (inserts) for BW Vul positioned along a logarithmic wavelength axis (nm). Note the strongly decreasing amplitude of variation A and the changing structure of the stillstand towards infrared wavelengths. Based on data from Sterken *et al.* (1987).

of β Cep stars were made photoelectrically, and a number of broad-lined β Cep stars were subsequently found. Broadening of the lines in those stars is not necessarily a consequence of rotation only, since other atmospheric effects also contribute to the broadening.

Another selection effect surely is the fact that for many decades no β Cep star fainter than $V = 6\overset{\text{m}}{.}5$ was known. HD 80383, the first relatively faint β Cep star, was only discovered in 1977 (Haug 1979), and many even fainter β Cep

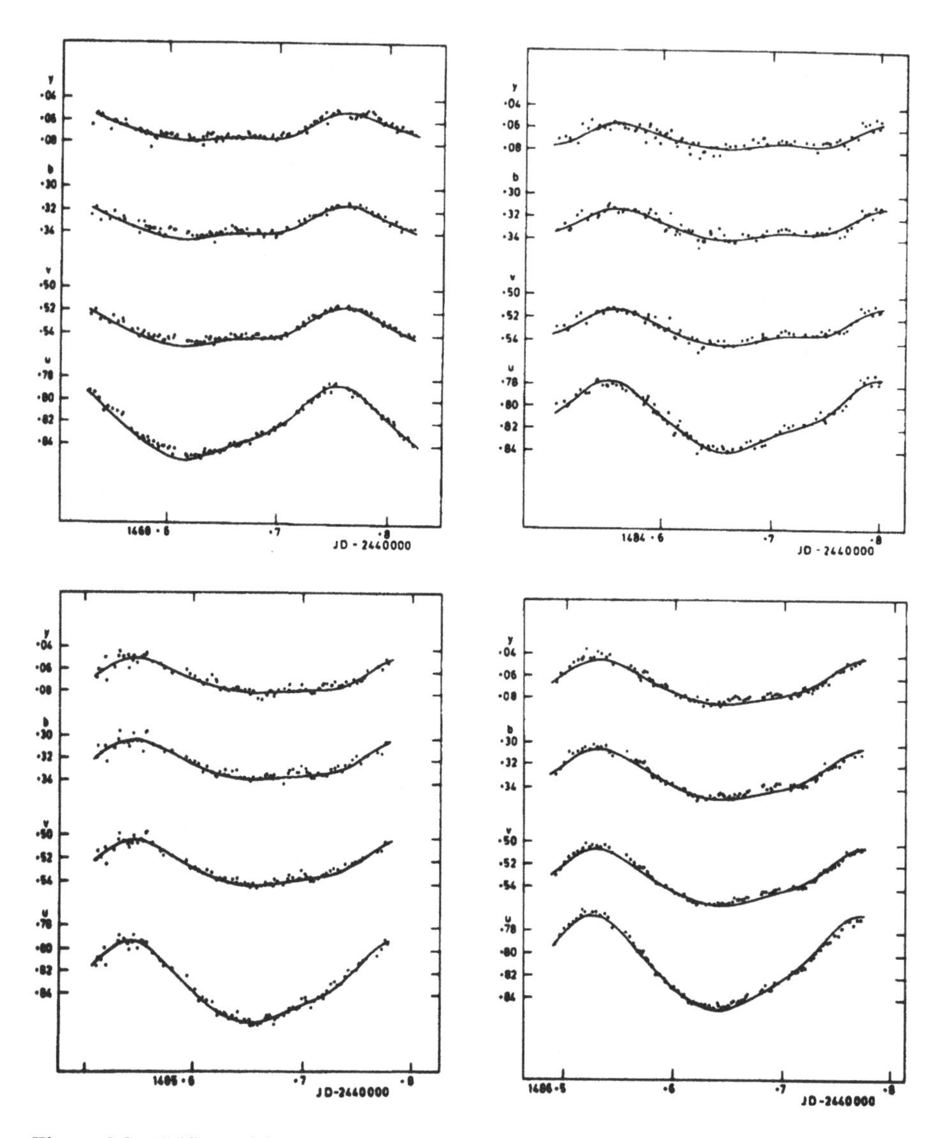

Figure 3.8 Differential u, v, b and y light curve of σ Sco = HR 6084 ($P = 0^{\mathrm{d}}.24684$). Comparison star is τ Sco (Jerzykiewicz & Sterken 1984).

stars were discovered later. Several systematic programs aimed at discovering β Cep stars have been carried out later on. Only two β Cep stars, HD 92024 (see Fig. 3.6) and 16 Lac, are known to be members of eclipsing pairs. No β Cep stars have so far been found in external galaxies, and any systematic searches for such stars have been confined to the LMC and SMC.

The small range of spectral types and luminosities among the β Cep stars make them cluster in a small region in the H–R diagram. This region is

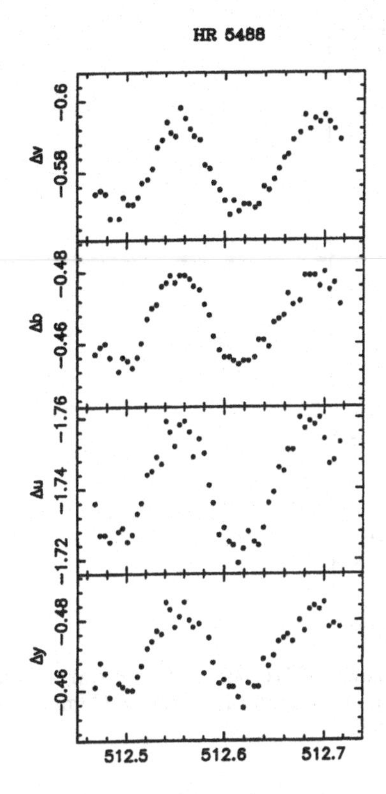

Figure 3.9 Differential u, v, b and y light curve of HR 5488 (β Cep star, $P = 0\overset{d}{.}1275504$). Comparison star is HD 130572 (vander Linden & Sterken 1985).

commonly labeled the 'β Cep instability strip'. It should be mentioned, however, that some Be stars share this area in the H–R diagram with the β Cep stars. Also, it should be stressed that stars seen as β Cep stars at one time may become Be stars at another (a most notorious example is β Cep itself, showing unprecedented strong emission in the core of the Hα line in 1990, Mathias *et al.* 1991). And, conversely, one case is known of β Cep pulsation appearing in a well-observed Be star: 27 EW CMa developed a pulsation somewhere between 1987 and 1990 (Balona & Rozowski 1991).

BW Vul (B2III, V = 6.55) has the largest known amplitude of light variation and radial velocity variation among the β Cep stars. The light curve is marked by a stillstand phase, the beginning of which precedes time of maximum by about $0\overset{d}{.}05$, the duration of the stillstand is close to $0\overset{d}{.}03$. The peak-to-peak amplitude of the light variation is approximately $0\overset{m}{.}2$ in the visual domain and increases to about $1\overset{m}{.}2$ at ultraviolet wavelengths (see Fig. 3.7); the period of the variation is approximately 5^{h} and is now secularly increasing at a rate of about 2 seconds per century.

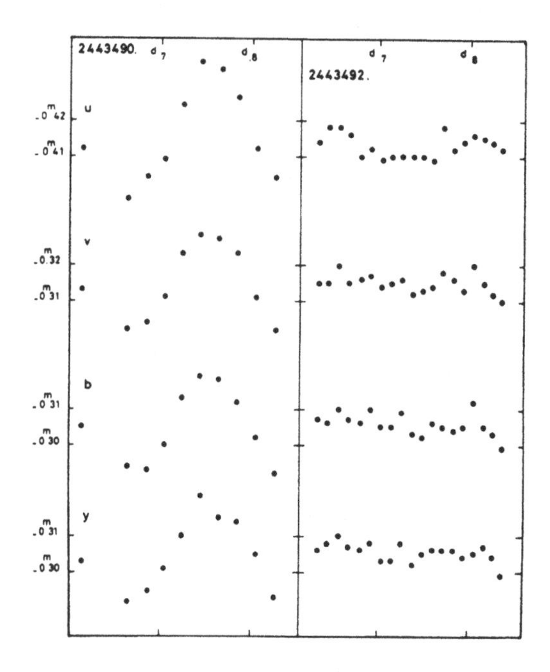

Figure 3.10 Differential u, v, b and y light curve of HD 64365 (β Cep star) on two nights when the star showed the maximum and minimum light range. Comparison star is HD 64287 (Jerzykiewicz & Sterken 1979).

In 1966 Christy pointed out that the region of ionisation of He^+, responsible for destabilising classical Cepheids (Section 3.7), δ Scuti (Section 3.5), RV Tau (Section 3.9) and RR Lyrae variables (Section 3.6), cannot be effective in β Cep stars (Christy 1966). Only very recently, the problem has been solved by invoking the classical κ-mechanism (Baker & Kippenhahn 1962), acting in a zone with temperatures of about 200 000 K. Iglesias *et al.* (1987), at the Lawrence Livermore National Laboratory, have developed a new opacity code, OPAL. On the basis of this code, Iglesias *et al.* (1990, 1991, 1992) presented improved opacity tables for Cepheid models, and showed that effects of spin-orbit interactions significantly enhance opacity in the region critical for driving pulsations of β Cep stars. Dziembowski & Pamyatnykh (1993) and, independently, Gautschy & Saio (1993), forced a breakthrough by showing that the application of the newest OPAL opacities for β Cep models explains the observed (fundamental and overtone) pulsation modes.

Illustrative examples of typical β Cep stars are σ Sco (Fig. 3.8), HR 5488 (Fig. 3.9) and HD 64365 (Fig. 3.10). For further reading, we refer to Lesh & Aizenman (1978) and Sterken & Jerzykiewicz (1994).

3.3 Be stars

C. Sterken

The first star in which the Be phenomenon – originally, that was the appearance of one emission line at one epoch in the spectrum of a star – was observed was γ Cas (Secchi 1866). Be stars are often called 'γ Cas' stars, or 'λ Eri' stars if they are periodic (see Balona 1991). γ Cas is probably the best-studied Be star,[1] but nowadays the term Be stars is almost universally used for naming the group of O6–B9 stars with luminosity class V to III that show (variable) emission in the Balmer lines associated with a rapidly-rotating circumstellar envelope or shell.[2] Baade (1992) narrows this definition by including a fast highly-ionised wind at higher altitudes, fast rotation, and a ratio V/R of the equivalent width of the violet and red emission components that is variable, but about unity on the average.

The Be stars share approximately the same area in the H–R diagram as the β Cep and 53 Per/mid-B stars (see Sections 3.2 and 3.4). Again, it should be stressed that stars seen as β Cep stars at one time may become Be stars at another (a most notorious example is β Cep itself, showing unprecedented strong emission in the core of the Hα line in 1990, Mathias *et al.* 1991). And, conversely, one case is known of β Cep pulsation appearing in a well-observed Be star: 27 EW CMa developed a pulsation somewhere between 1987 and 1990 (Balona & Rozowski 1991). 27 CMa also showed fading events – that is, sudden fadings of a few weeks duration, they were preceded by quasi-periodic oscillations on time scales of 10–20 days (Mennickent *et al.* 1994).

Many Be stars display short or intermediate-term light variability (Percy 1987, Cuypers *et al.* 1989, Cuypers 1991, Balona *et al.* 1987, 1992, see also Baade 1987, Balona 1990 and Štefl *et al.* 1995). The periods are from about $0\overset{d}{.}4$ to 3 days, multiple periods and double waves are found, and the amplitudes are in the range $0\overset{m}{.}01$ to $0\overset{m}{.}3$. Emission line strength and brightness are sometimes correlated, a very interesting example is the inverse correlation displayed in Fig. 3.11 for γ Cas and V744 Her. A bi-periodic star with unusually large periods ($P_1 = 8\overset{d}{.}929, P_2 = 2\overset{d}{.}824$ and V amplitude $0\overset{m}{.}8$) is HD 137518 (see Fig. 3.12).

Harmanec (1983, 1994) points out that the observed long-term variations of Be stars can, in principle, be classified into two categories, viz. light curves reminiscent of mild cases of nova outburst (e.g. γ Cas), and light curves which are the inverse of nova light curves, thus reminiscent of R CrB stars (see

1. γ Cas and λ Eri are the only stars to have been reported to be X-ray Be stars (Smith 1994b).
2. Jaschek *et al.* (1981) use the more general concept 'a non-supergiant B-type star whose spectrum has or had at some time one or more hydrogen lines in emission'.

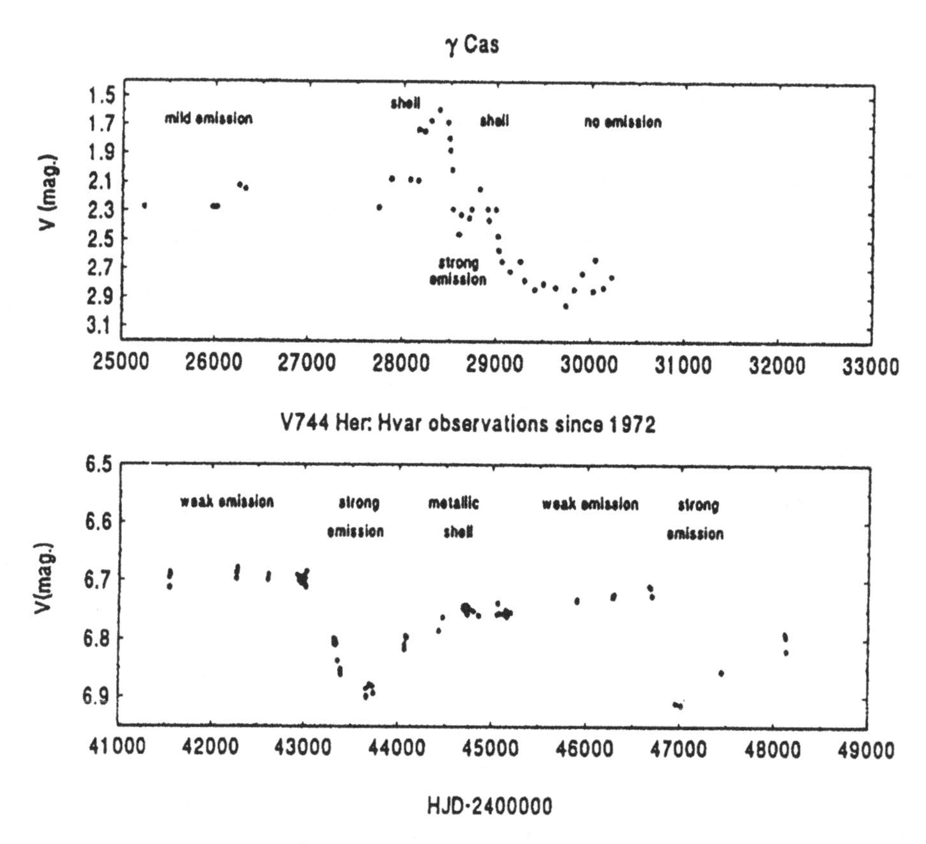

Figure 3.11 Long-term light variation of γ Cas and V 744 Her, illustrating typical examples of positive and inverse correlation, respectively, between the emission strength and brightness (Harmanec 1994).

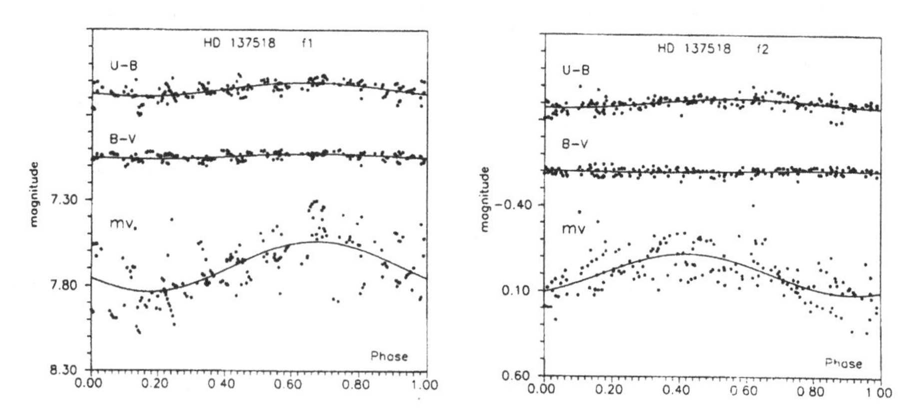

Figure 3.12 m_V, $U-B, B-V$ phase diagrams for the large-amplitude bi-periodic star HD 137518 after prewhitening for the first frequency (corresponding to $P_1 = 8^d\!.929$) and the second frequency (corresponding to $P_2 = 2^d\!.824$) (Waelkens 1991a).

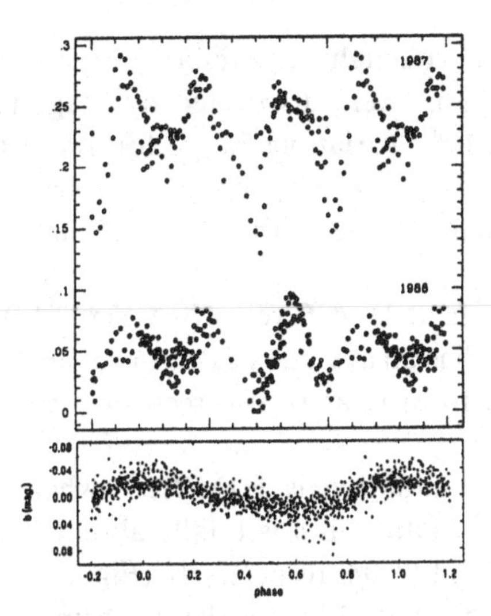

Figure 3.13 Differential *b* phase diagram of η Cen ($P = 1^{\mathrm{d}}.927$) showing the very uncommon triple-wave light curve in 1987 and 1988 (Cuypers *et al.* 1989) (top), and also the single-wave $P = 0^{\mathrm{d}}.6424241$ solution presented by Štefl *et al.* (1995) (bottom).

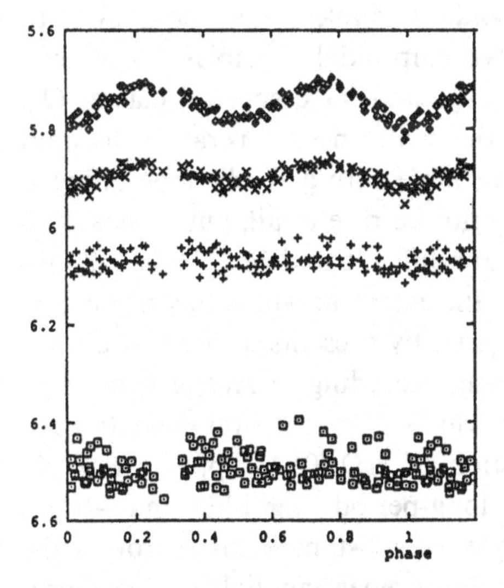

Figure 3.14 *uvby* phase diagram of HD 50123, an interacting B6 Ve + K0 III binary with $P = 28^{\mathrm{d}}.601$ (Sterken *et al.* 1994). Note the progression in amplitude from *u* (bottom) to *y* (top).

Section 2.5), like BU Tau and V 477 Her. Balona (1990) has shown that the period of the line-profile and light variations in Be stars is not significantly different from the period of rotation of the star, and thus argues in favour of a rotational modulation (RM) mechanism rather than non-radial pulsation (NRP) as the cause of these variations.

η Cen is a most enigmatic Be star. Its light variability was discovered by Cousins (1951), and its rapid variations were discovered by Stagg (1987). Cuypers *et al.* (1989) reported periodic light variations in the form of a triple-wave light curve with $P = 1\overset{\mathrm{d}}{.}927$; Štefl *et al.* (1995), however, described the photometric data quite well with a single wave with one dominant period $P = 0\overset{\mathrm{d}}{.}642424$ and its overtones – this conclusion was also supported by spectroscopy, and is secularly stable over at least 6 years (see Fig. 3.13). The light curve is slightly non-sinusoidal and has an average peak-to-peak amplitude of $0\overset{\mathrm{m}}{.}05$ in y, b and v, increasing to $0\overset{\mathrm{m}}{.}10$ in u, as is also seen in β Cep stars (compare, for example, with Fig. 3.7).

A case with the completely opposite situation is HD 50123, where the v amplitude is extremely small, and any variation in u is totally absent. b and y amplitudes, however, amount to $0\overset{\mathrm{m}}{.}05$ and $0\overset{\mathrm{m}}{.}08$, respectively (Fig. 3.14). This peculiar behaviour is explained by Sterken *et al.* (1994) as the star being a B6 Ve + K0 III interacting binary where the rotation of the elliptically deformed Be star due to tidal distortion of the Roche-lobe filling late-type companion causes the dominant light variations. HD 50123 is certainly an outstanding star, with reference as well to the class of ellipsoidal variables, as to the class of Be stars. Beech (1985) lists a total of 27 known ellipsoidal variables, 26 of them have orbital periods between 0.8 and 5.6 days and have spectral classes O to G2; they are mostly dwarfs. HD 50123, with its $28\overset{\mathrm{d}}{.}6$ period, is rather dissimilar, and it is also the first known ellipsoidal variable with a Roche-lobe filling K giant component. Stars like HD 50123 may not be rare at all, but – because of their small amplitudes and long periods – are difficult to detect. HD 50123 may also be a keystone in the controversy over the extent to which binarity is important to explain the Be phenomenon: it apparently presents a case just at the binary period limit $\approx 10^{-1}$ years, below which, according to Abt & Cardona (1984), no Be binaries should exist (a view which, however, was heavily questioned by Harmanec 1987). The configuration of HD 50123 appears typical of the so-called W Ser stars[1] (a group of long-period Algol-like mass-transferring binaries), which are characterised by very substantial discs around the more massive components, strange and poorly-repeating light curves, prominent optical emission lines and large secular period changes (Wilson 1989) – typical members are RX Cas, SX Cas, W Ser, W Cru and β Lyr, and are classified in *GCVS* as EA/GS or EB/GS, see Table 1.6. In particular HD 50123 resembles the strongly interactive eclipsing binary SX Cas (B7 + K3 III), viewed at a lower inclination so that it does not eclipse, and the accretion disc does not obscure the primary B-type star.

1. The W Serpentis group was defined by Plavec(1980), but is not part of the *GCVS* classification scheme.

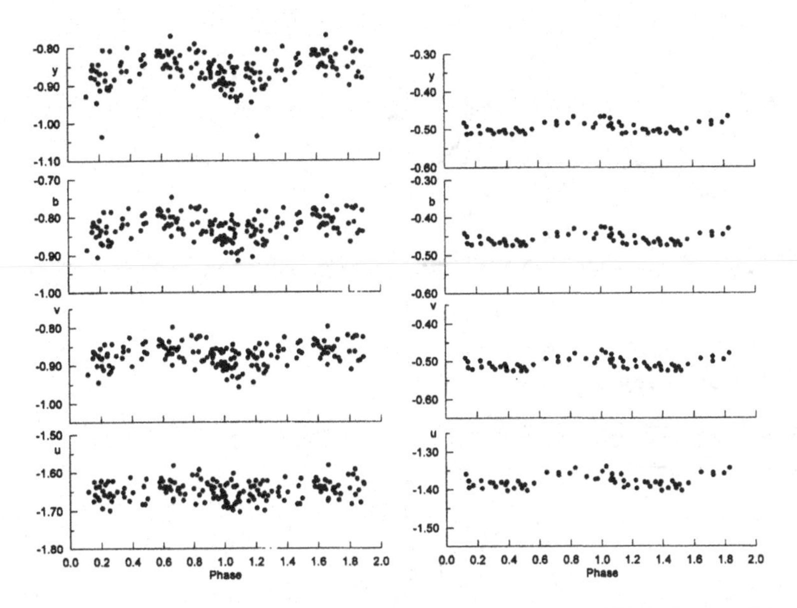

Figure 3.15 *uvby* phase diagrams for HR 2492, $P = 87^{\rm d}\!.9$ (left) and HR 2855, $P = 92^{\rm d}\!.7$ (right) (Sterken *et al.* 1996).

An interesting case of Be star variability was reported by Sterken & Manfroid (1996) for HR 2517. For about 10 years the star exhibited signs of microvariability in *uvby*, but during about the last two years recurrent flash-like brightening over more than a tenth of a magnitude in the *y*, *b* and *v* bands, with additional flare activity in *u* has been observed. The photometric behaviour of HR 2517 shows a striking resemblence (though on a somewhat slower time scale) to that observed by Balona (1990) in κ CMa: a multiple-peak outburst amounting to $0^{\rm m}\!.1$ suddenly appears, and fades over several days (see Fig. 3.16). Balona interprets this phenomenon as due to a bright area suddenly appearing and developing, thus ruling out the explanation of Be star light variability by non-radial pulsation alone. As the star rotates, the modulation in photospheric temperature produces periodic light and profile variations. Sterken & Manfroid (1996) suggest an alternative explanation for the flaring in terms of the possibility that HR 2517 is an eccentric, massive close binary system, where the primary has lost a substantial amount of mass so that the secondary obtains an internal structure very similar to that of a single star of comparable mass. A subsequent supernova explosion could then make a compact remnant orbit the mass gainer, and a (transient) HMXRB (see Chapter 7) is observed. In such HMXRB configurations where the secondary star is a Be star, a transient X-ray source is often found, and this is explained by the fact that a detectable X-ray luminosity is only expected when

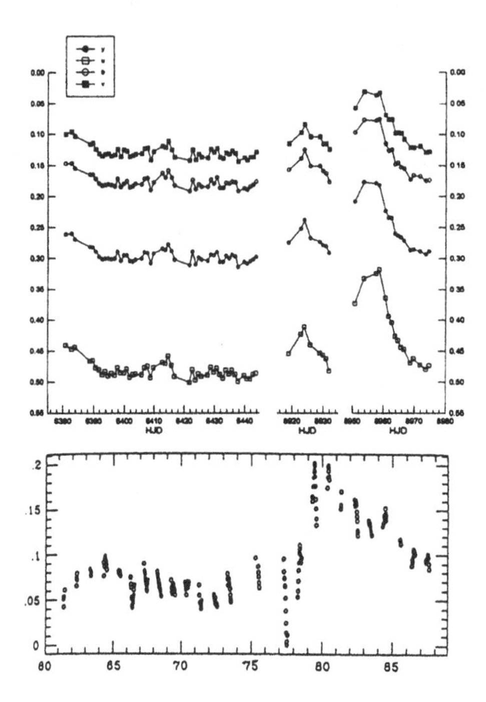

Figure 3.16 Flares observed in HR 2517 (Sterken & Manfroid 1996, *uvby* photometry) and κ CMa (Balona 1990). The X-axis of the bottom curve is HJD – 2 446 700.

the compact star crosses the dense equatorial disc of the Be star (Kaper 1994, van den Heuvel & Rappaport 1987).

For more general reading on Be stars, we recommend Slettebak (1979). Specific information about variability of Be stars on short time scales (a couple of hours to a couple of days) can be found in Harmanec (1989), Gies (1991) and Baade (1992). Papers related to NRP explanations were published by Smith (1977), Baade (1987), Bolton (1981) and Vogt & Penrod (1983); RM-related papers are by Harmanec (1983, 1989) and Balona (1990, 1991), see also Baade & Balona (1994). A most interesting paper dealing with these matters is Štefl *et al.* (1995).

3.4 53 Per/mid-B/slowly-pulsating B variables

C. Sterken

The concept '53 Per' variables was introduced by Buta & Smith (1979) and Smith (1980) for O8 to B5 stars that show variability in line profiles with variable periods of the order of a day, thus far too long and too unstable to be

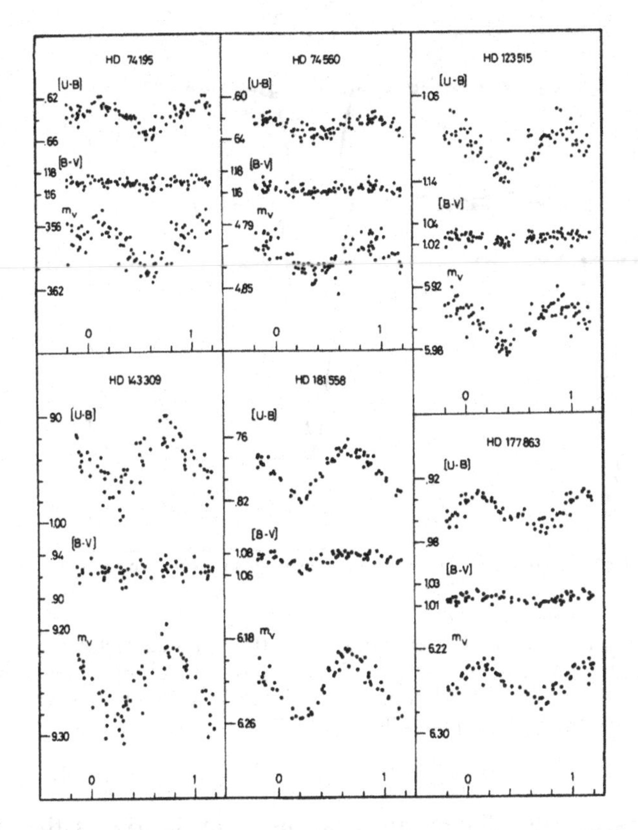

Figure 3.17 $U - B, B - V$ phase diagrams for six mid-B variables (Waelkens & Rufener 1985) .

associated with β Cep-type variability (Section 3.2). Smith classified the 53 Per variables as non-radial pulsators.

Waelkens & Rufener (1985) introduced the class of 'mid-B' variables, B3–B8 stars of luminosity class III–V; they have periods of 1–3 days and amplitudes of light variations of a few $0\overset{\mathrm{m}}{.}01$; the colour variations are in phase with the light variations, and the colour-to-light ratio remains constant despite variability in amplitude on a cycle-to-cycle and even on a year-to-year base (see Fig. 3.17). They inferred that the 53 Per class and the mid-B class of variables are identical, a conclusion supported by subsequent line-profile work (for more details, see Cuypers 1991). Later, Waelkens (1991b) reports that his observations of all mid-B stars reveal that all these stars are multiperiodic variables with periods in the range between 1 and 4 days, a multiperiodicity that points to non-radial pulsation as the explanation for the variability of these stars – an explanation that is also supported by the line-profile variations which are consistent with low-degree non-radial pulsation modes. Consequently, he renames these stars

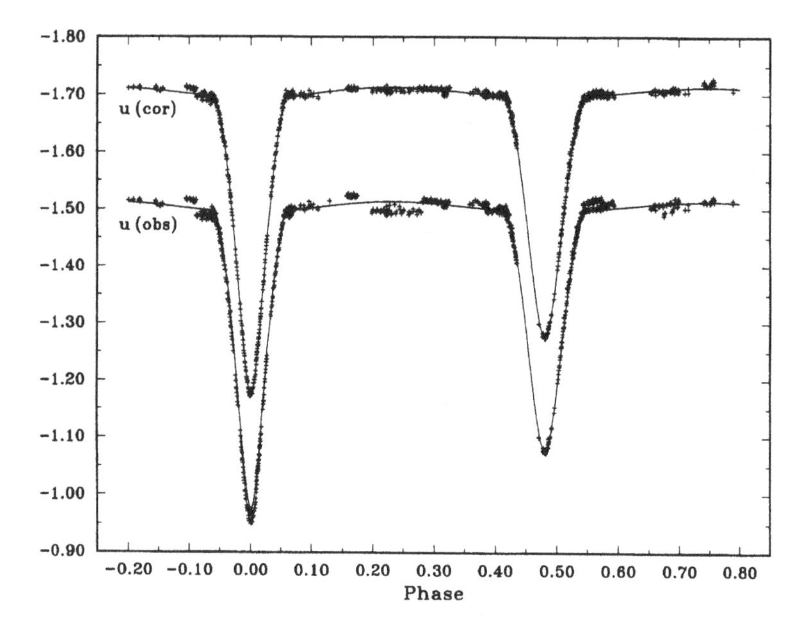

Figure 3.18 V 539 Ara: observed (lower, shifted by $+0\overset{\mathrm{m}}{.}2$) and corrected for pulsational variability *u* phase diagram; the full line represents the model solution to the corrected data (Clausen 1996). Eclipses have depths of $0\overset{\mathrm{m}}{.}5$ and $0\overset{\mathrm{m}}{.}4$ and the secondary eclipse is displaced from phase 0.50.

'slowly pulsating B stars' (SPBs). The identification of pulsation as the cause of the slow variability of the variable mid-B stars is supported by theoretical calculations by Gautschy & Saio (1993) and Dziembowski & Pamyatnykh (1993).

So far, the nomenclature on these stars is somewhat confusing. '53 Per' refers to the prototype of the group of (pulsating) mid-B variables, whereas 'SPB' refers to the same kind of variables, but with a nomenclature that involves a physical mechanism. The trouble is that the northern-hemisphere observers came to the discovery of this class of variables through spectroscopic observations, while the southern observers arrived at delineating an instability region via photometric methods. The situation, in fact, bears some resemblance to β Cep variables, which were first discovered by spectroscopic means, eventually leading to the extension of the period range (to short periods) via photometric measurements.

Balona & Cuypers (1993) found HR 2680, a B5 V + A–F single-lined spectroscopic binary to be a slowly pulsating B star with variable period of oscillation and an amplitude of $0\overset{\mathrm{m}}{.}03$. Clausen (1996) made the very important discovery that the secondary component of the double-lined eclipsing binary V 539 Ara (Fig. 3.18) is a slowly pulsating B star with $P_1 = 1\overset{\mathrm{d}}{.}36$ and $P_2 = 1\overset{\mathrm{d}}{.}78$. V 539

Ara is one of the few double-lined eclipsing binaries on which our knowledge of accurate masses of B stars is founded (Andersen 1991, Popper & Hill 1991), and the combination of well-established SPB oscillations and accurate dimensions is unique.

3.5 δ Scuti variables

M.W. Feast

The δ Scuti variables are pulsating variables with periods less than 0.3 d, spectral types A or F, and visual light amplitudes in the range from a few thousandths of a magnitude to about 0.8 mag. A magnitude range of about 0.2 mag is typical for the δ Scuti stars at present known. The stars form a group which lies in an instability strip in the H–R diagram which includes the classical Cepheids at its bright end and the pulsating white dwarfs at its faintest limit.

There has been a great deal of confusion over the nomenclature applicable to these stars. This has been particularly troublesome since the broad class contains stars belonging to both halo and young disc populations in our Galaxy. Alternative names for all or some of these variables are: dwarf Cepheids, RRs variables, AI Vel stars, SX Phe stars, and ultra-short period Cepheids.

δ Scuti stars can show very complex light variations since, while some of them are pulsating in one radial mode only, others may be pulsating simultaneously

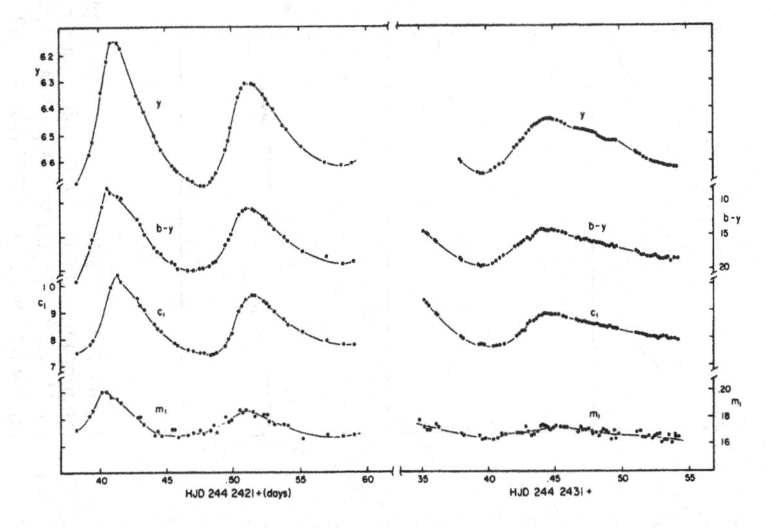

Figure 3.19 $y, b-y, m_1, c_1$ light curve of AI Vel (δ Scuti) (Breger 1977). The curves have been drawn freehand.

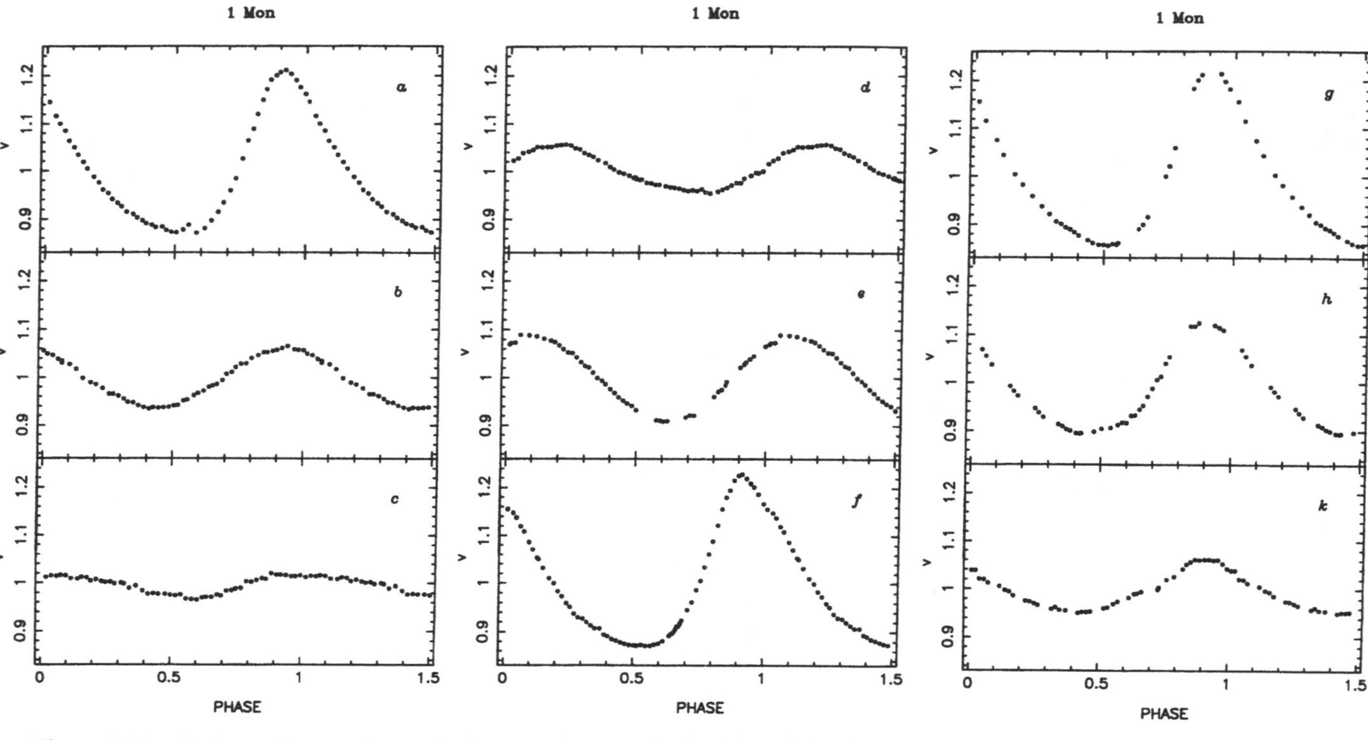

Figure 3.20 V phase diagram (natural photometric system) of 1 Mon (δ Scuti) according to ephemeris 2441681.7233 + 0.13612E (Shobbrook & Stobie 1974). Julian date range: (a) 244 1661.1123–1661.2443 (b) 244 1662.9485–1663.2436 (c) 244 1663.9353–1664.2394 (d) 244 1664.9468–1665.1399 (e) 244 1665.9404–1666.2309 (f) 244 1668.1159–1668.2491 (g) 244 1668.9333–1669.2373 (h) 244 1669.9386–1670.2448 (k) 244 1670.9335–1671.1773.

in several radial and non-radial modes and it is possible that in some cases mode-switching takes place. A detailed study of these complexities can yield important information on the star's internal structure. Such work is classified generally as asteroseismology. Unno *et al.* (1988) give an overall discussion of the problem of nonradial oscillations in stars.

The visual absolute magnitudes of these variables range from about +3.0 to 0.0. The *GCVS* lists over 200 members of this and closely related classes (e.g. SX Phe stars) but, especially in view of the very low visual amplitudes of many δ Scuti variables, it is clear that many more remain to be discovered even amongst the relatively bright stars.

The figures demonstrate some of the complexities of these stars. AI Vel (Fig. 3.19) shows four or more interfering periods. Very extensive observations have been made of this star over many years by Walraven and collaborators (Walraven 1955, Walraven *et al.* 1992). Figure 3.19 shows the light curves over a period of about 5 hours at two different times. The nine V light curves of 1 Mon in Fig. 3.20 were obtained during a ten day period and each night's run of about 5 hours is shown separately. The phases are calculated with a period of $0\overset{d}{.}13612$.

3.6 RR Lyrae variables

M.W. Feast

These stars were, in the past, often referred to as cluster-type variables because they occur in considerable numbers in globular clusters and this designation is still sometimes used. A few galactic globular clusters have over one hundred RR Lyrae members but others have only a few or none at all. Variables similar to those in clusters are found in the general field and are members of the galactic halo and old disc (and/or thick disc) populations.

RR Lyrae stars are radial pulsators with periods in the approximate range 0.2 to 1.0 day. Their metallicities span a wide range, from about the solar value to a hundred times less. As with the Type II Cepheids (Section 3.8), which are probably a different evolutionary stage of the same kind of star, their spectra show evidence of shock waves being propagated outwards through their atmospheres once per cycle.

On the basis of their light curves the RR Lyrae can be divided into two main groups. The RRab stars have relatively high light amplitudes (visual amplitudes of about one magnitude are common) and asymmetrical light curves (a steep rising branch). These stars are believed to be pulsating in their fundamental mode. The RRc variables have lower light amplitudes (perhaps about $0\overset{m}{.}5$)

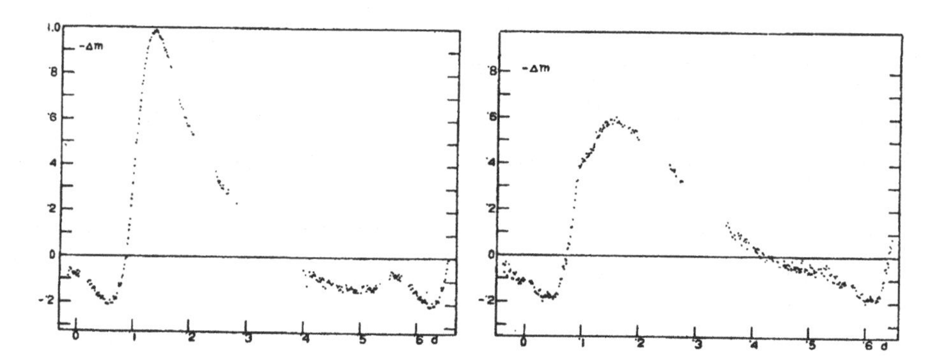

Figure 3.21 RR Lyr at two phases of its Blazhko cycle (Fig. 1 of Walraven 1949). Variation in magnitude plotted against time (0.0 to 0.6 days).

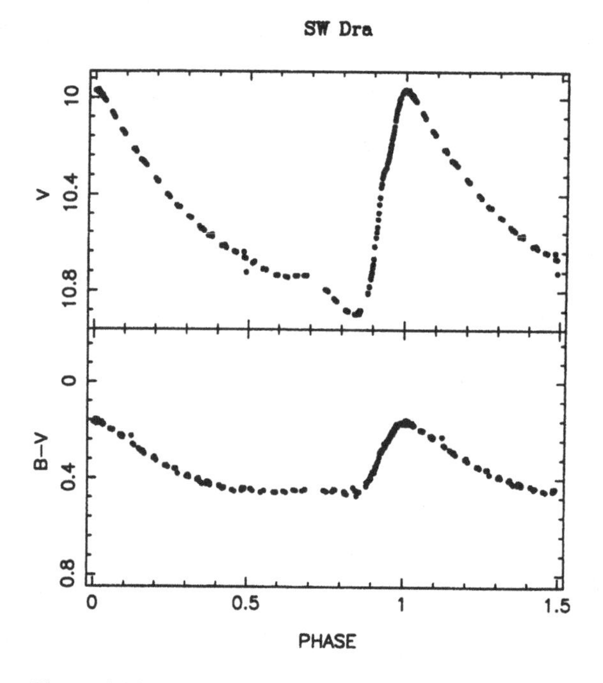

Figure 3.22 $V, B - V$ phase diagram of SW Dra (RRab) according to ephemeris $2446495.7493 + 0.56966993E$ (Jones *et al.* 1987).

and more nearly sinusoidal light curves. They are believed to be pulsating in their first overtone. The periods of RRab variables are mostly in the range 0.4 to 1.0 day and the RRc in a range from about 0.2 to 0.5 day. In a given globular cluster the amplitudes of the RRab variables decrease as one goes to variables of longer period.

As in the case of classical Cepheids (Section 3.7), a few RR Lyrae stars show variations in the shape and amplitude of their light curves from cycle

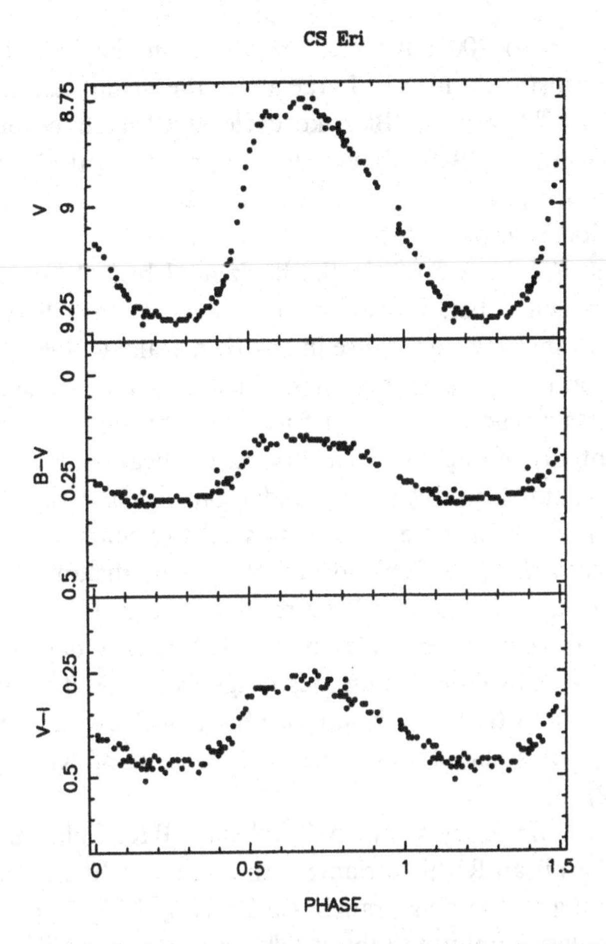

Figure 3.23 $V, B-V, V-I$ phase diagram of CS Eri = HD 16456 (RRc) according to ephemeris $2440000.0 + 0.311331E$ (Dean *et al.* 1977).

to cycle due to the simultaneous excitation of two periods, for example AQ Leo (Jerzykiewicz & Wenzel 1977). In the case of the RR Lyrae stars, the two periods which are beating together are the fundamental and first overtone. In recent literature these double-mode RR Lyraes are often called RRd stars although the *GCVS* refers to them as RRb variables. The ratio of the first overtone to the fundamental period is about 0.746. One triple-mode RR Lyrae (AC And) is known: it is pulsating in the first and second overtones as well as in the fundamental. No RR Lyraes have yet been identified with certainty as pulsating only in the second overtone.

Many RR Lyrae variables show long-term modulations of their light curves. RR Lyrae itself is one of these. This phenomenon is known as the Blazhko effect (see Fig. 3.21). Its cause is still not understood. The modulation periods are generally in the range 20 to 200 days and the effect on the light curve can be quite marked. For instance, in RR Lyrae itself the visual amplitude of the star varies by about $0^{m}.3$ over the Blazhko cycle and the shape of the light curve changes (see Walraven 1949, Teays 1993). For some variables the Blazhko effect itself is modulated on an even longer time scale. In the case of RR Lyrae the tertiary period is approximately 3.8 to 4.8 years.

In the H–R diagram the RR Lyraes fall in the horizontal branch region, a stage of stellar evolution which is believed to occur between a star's first and second ascent of the giant branch. Besides those in our own Galaxy, RR Lyraes have been detected in several extragalactic systems including the Magellanic Clouds, the Andromeda galaxy and several dwarf members of the local group of galaxies. They are potentially of importance as distance indicators, especially in old stellar systems (for example in dwarf spheroidal galaxies and the Bulge region of our own Galaxy) in which there are no classical Cepheids to use for distance estimates. A concern that the Cepheid and RR Lyrae distance scales were distinctly different now appears to have been resolved (Feast 1995). There is good evidence that the absolute magnitudes of the RR Lyraes depend on their metallicities. Their visual absolute magnitudes range from about +1.0 for stars with metallicities similar to that of the sun, to about +0.5 for stars with metallicities one hundredth of solar. For a general discussion of RR Lyrae variables see Nemec (1992).

The figures illustrate the differences between RRab and RRc light curves. SW Dra (Fig. 3.22, $P = 0^{d}.56$) is an RRab variable with a visual light amplitude of nearly one magnitude and a steep rising branch. CS Eri (Fig. 3.23, $P = 0^{d}.31$), an RRc variable, is of smaller amplitude (about $0^{m}.5$) and has a much more nearly sinusoidal light curve.

3.7 Cepheid variables

M.W. Feast

In mid-1784 only five variable stars (apart from novae and supernovae) were known. Four of these were what we now call Mira variables and one was an eclipsing star (Algol). Then, on September 10, 1784, Edward Piggot established the variability of η Aquilae whilst his friend John Goodricke showed that β Lyrae was variable. Shortly afterwards Goodricke found δ Cephei to vary. Whilst β Lyrae is the prototype of an important class of eclipsing variables (see Section 6.1), δ Cep and η Aql are what we now call Cepheids and with periods of $5^{d}.4$ and $7^{d}.2$, respectively, and visual light amplitudes of about $0^{m}.9$, they are fairly representative of this class. The term Cepheid was at one time applied generally to any continuously varying star with a regular light curve and a period less than about 35 days, unless it was known to be an eclipsing star. It is now recognized that the class defined in this way is heterogeneous, containing stars in different mass ranges and evolutionary states. Stars with periods less than one day are now treated separately (they are mainly RR Lyrae variables, see Section 3.6). Type II Cepheids and RV Tauri stars are also treated separately and their distinguishing features are dealt with in, respectively, Sections 3.8 and 3.9. The remaining stars are called δ Cephei variables, Type I Cepheids, classical Cepheids, or simply (and most frequently), Cepheids. Eggen (1970) has used the term Cepheid for a wider variety of stars, subdividing them into long-period, short-period, very short-period, ultra short-period and pseudo-Cepheids. But this terminology has not been generally adopted. In 1914 Shapley considered the then known properties of Cepheids and concluded that they were most likely pulsating variables. However he seems to have thought of them as non-radial pulsators and it was left for Eddington to deduce that they pulsate radially. Detailed studies of their light, colour and radial velocity variations leave no doubt that this interpretation is correct.

Cepheids are strictly periodic variables with periods ranging from about 1 day to about 50 days and with a few extreme examples up to 200 days. The general form of the light curve varies smoothly as one moves from shorter to longer period stars. This is known as the Hertzsprung progression after the Danish astronomer who investigated it. The shorter period variables have steep, narrow maxima. With increasing period the relative widths of the maxima broaden. At periods of around 8 to 10 days the maxima often appear double. At shorter periods than this there are frequently bumps on the falling branch, whilst at longer periods the bumps tend to occur on the rising branch. The longer period stars (20 to 40 days) generally have very steep rising branches but the light curves of the longest period Cepheids are more nearly sinusoidal. Fourier decomposition of the light curves allows one to study the Hertzsprung progression quantitatively (see for example, Simon 1988). It has been known since the work of Stobie in 1969 that comparison of the phase at which the bump occurs with pulsation theory can lead to an estimate of the stellar mass but it is only with the recent major revision in the calculation of the opacity of stellar material that some measure of agreement has been reached between masses determined in this way and other estimates (see Moskalik *et al.* 1992). The bump phenomenon is due to a resonance between the fundamental pulsation mode and the second overtone.

Early in this century Henrietta Leavitt discovered one of the most useful relations in astronomy when she noticed that the magnitudes of Cepheids in

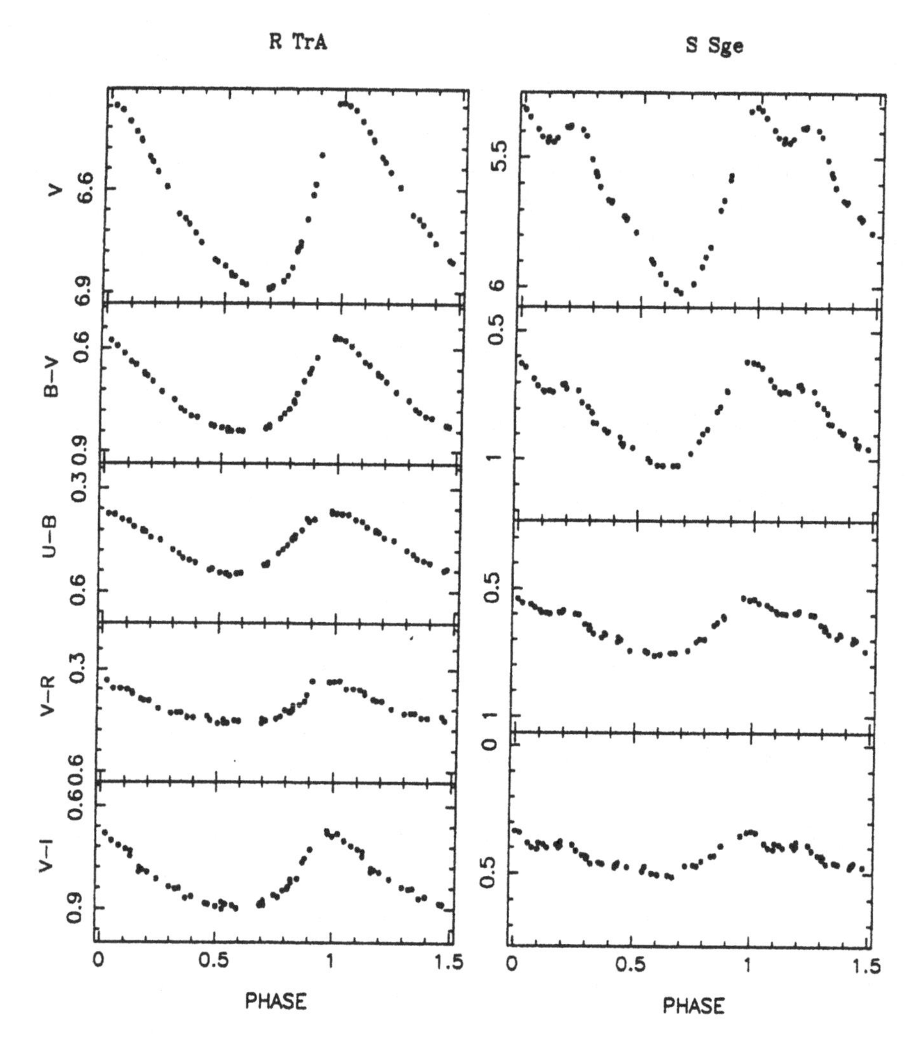

Figure 3.24 $V, B-V, U-B, V-R, V-I$ phase diagram of R TrA and S Sge (Cepheids). Left: R TrA, ephemeris 2440838.21 + 3.389287E (Pel 1976, data from Gieren 1981), note the narrow maximum compared to the minimum. Right: S Sge, ephemeris 2435688.250 + 8.382173E (Moffett & Barnes 1980), note the double maximum.

the Small Magellanic Cloud were related almost linearly to $\log P$ (brightness increasing with period). Modern work on Magellanic Cloud Cepheids has shown that there is an intrinsic scatter to this period–luminosity relation and both theory and observation show that the exact relation involves period, luminosity and surface temperature. In practice a colour index is used in place of temperature. These PL and PLC relations can be calibrated using Cepheids in young clusters in our own Galaxy since the distances of these clusters can be determined. Because Cepheids are bright objects, ranging in visual absolute

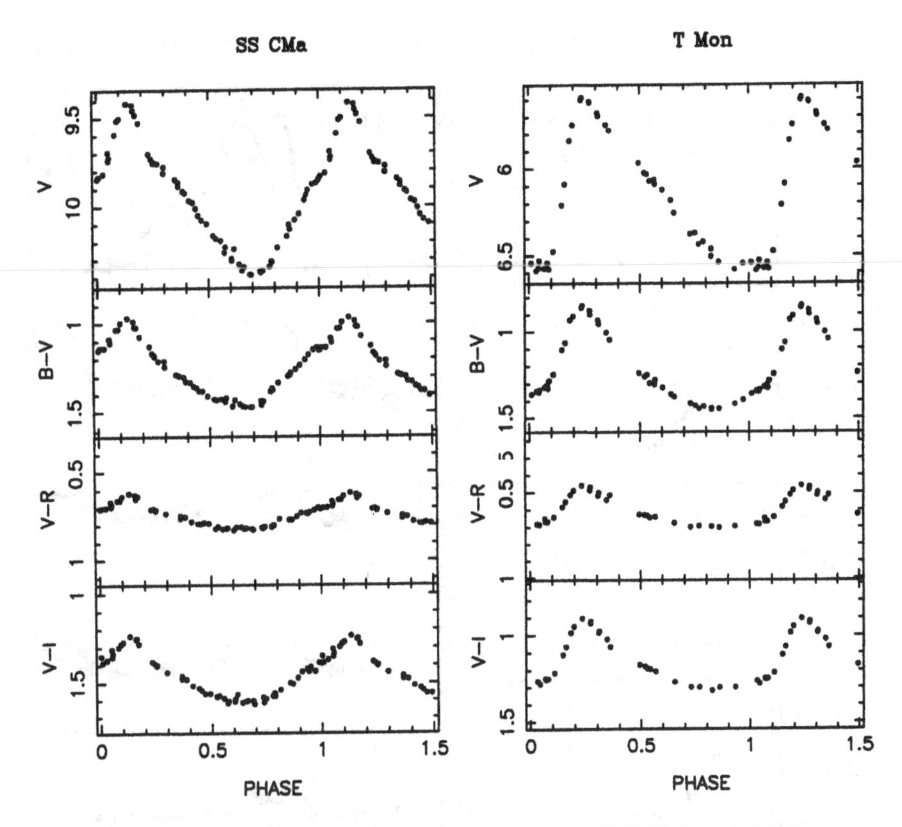

Figure 3.25 $V, B-V, V-R, V-I$ phase diagram of SS CMa and T Mon (Cepheids). Left: SS CMa, ephemeris 2420000.000 + 12.3580E (Caldwell & Coulson 1985), note the bumps on the rising and falling branches for this star near the transition period. Right: T Mon, ephemeris 2420000.00 + 27.0197E (Caldwell & Coulson 1985); the rising branch is very steep in V.

magnitude from –2 at a period of about 2 days to –6 near 40 days, they can be detected in galaxies as far away as the Virgo galaxy cluster using the Hubble space telescope (e.g. Freedman *et al.* 1994) and as such are the basis of the extragalactic distance scale.

Cepheids are relatively young, massive objects. Those with periods near 2 days have masses about five times that of the sun and ages of about 10^8 years. The mass increases and the age decreases with increasing period, being about 15 solar masses and 10^7 years at 40 days. Most known Cepheids pulsate in the fundamental mode but some pulsate in the first overtone. These latter variables have lower light amplitudes, nearly sinusoidal light curves and periods which are generally in the range 1.5 to 4 days. Because of their lower light amplitudes the overtone pulsators have tended to be overlooked in the past. It has only recently been recognized that these overtone pulsators rep-

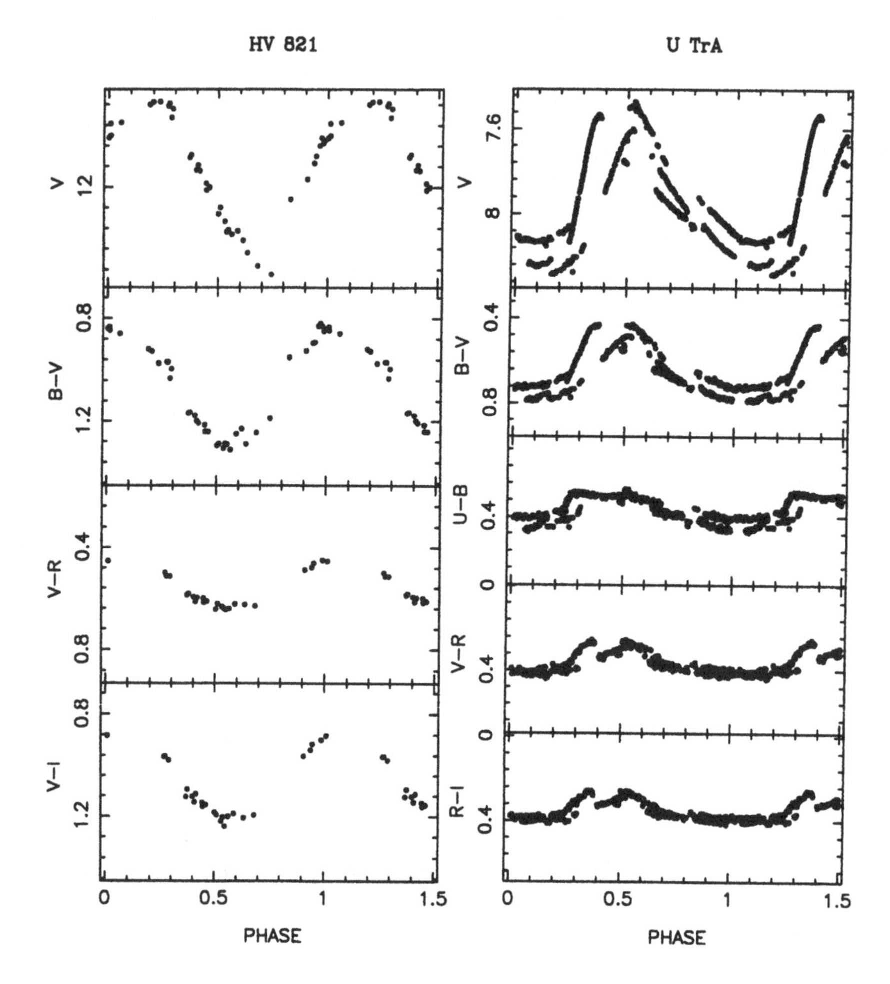

Figure 3.26 $V, B-V, V-R, V-I$ phase diagram of HV 821 (a very long period Cepheid in the Small Magellanic Cloud) and U TrA. Left: HV 821, ephemeris $2420000.0 + 127.6E$ (Caldwell & Coulson 1984); the light curve is more nearly sinusoidal than for shorter-period Cepheids. Right: U TrA, ephemeris $2443267.14 + 2.568423E$ (Faulkner & Shobbrook 1979); the first-overtone period is $1^{d}.82$. Note the varying light curve from cycle to cycle due to the beating of the two periods (a few representative cycles are overlaid).

resent a significant fraction of all Cepheids. For instance an analysis of the Cepheids found in the Large Magellanic Cloud by the EROS collaboration during a search for microlensing, shows that 30 percent of this sample are overtone pulsators (Beaulieu *et al.* 1995). Overtones can generally be quantitatively distinguished from fundamental mode pulsators by Fourier analysis of their light curves. A few Cepheids are probably pulsating in the second overtone.

In a small group of Cepheids, mainly with periods in the range 2 to 6 days, the shape and amplitude of the light and velocity curves vary markedly from cycle to cycle. It was recognized by Oosterhoff that this was due to the beating of two periods. Extensive data on double-mode Cepheids in the Large Magellanic Cloud have recently been obtained by the MACHO collaboration (Alcock *et al.* 1995). This work shows that there are two distinct classes of these stars, those pulsating in the fundamental and first overtone, and those pulsating in the fundamental and second overtone.

Evidence has been found for a gradual change of period in some Cepheids. Such changes are typically one second per year which is about the rate predicted by evolutionary theory. Cepheids have the characteristic spectral signatures of supergiants and spectral types F, G or K. Thus the surface temperatures of many of them are very roughly comparable to that of the sun.

The light curves have been chosen to show typical features at different periods. Extensive light and colour curves of Cepheids in the near infrared have also been published (e.g., Laney and Stobie 1992). In addition to the references cited above, the following books provide useful summaries: Cox (1980), Madore (1985).

3.8 Type II Cepheids

M.W. Feast

Type II Cepheids are probably best regarded as the low mass analogues of the (classical) Cepheids which are discussed in Section 3.7. A variety of designations have been used for these stars in the past. It is necessary to be aware of this especially in consulting the older literature. They are referred to as CW stars in the *GCVS* and a distinction is sometimes made between those with periods less than 8 days (which are then called BL Her or CWB stars, see also Fig. 3.27) and those with longer periods (called W Vir or CWA stars). Diethelm (1983) classified type II Cepheids in the 1 to 3 day period range into three groups, RRd stars (smooth light curves), CW (or W Vir) stars (a bump on the ascending branch), and BL Her stars (a bump on the descending branch). This nomenclature was later changed (Diethelm 1990) as follows: RRd to AHB1, CW to AHB2, and BL to AHB3. However it does not seem that either scheme has been widely adopted. It should be noted that the letters RRd are commonly used in the current literature to denote double mode RR Lyrae stars (see Section 3.6 and also Fig. 3.28. The complex and changing nomenclature reflects, at least in part, the uncertainty which still surrounds our understanding of

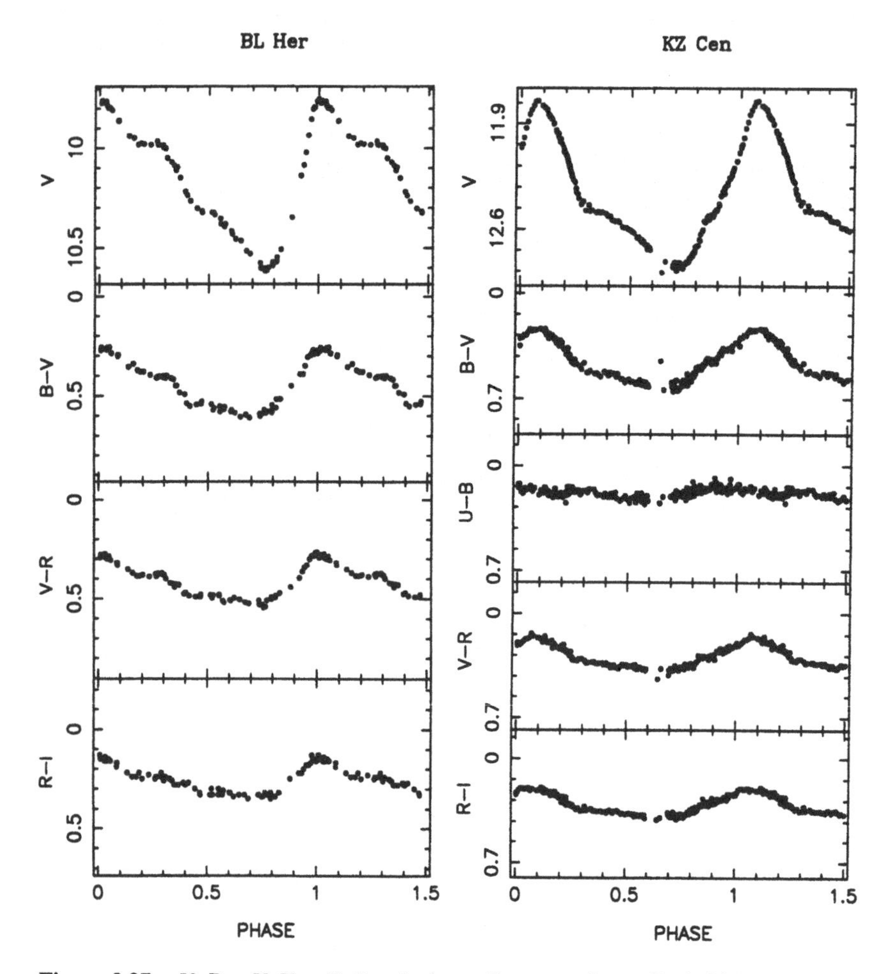

Figure 3.27 $V, B-V, V-R, R-I$ phase diagram of two Cepheid II stars. BL Her (ephemeris 2441841.289 + 1.307443E, Moffett & Barnes 1984), note the prominent bump on the descending branch. KZ Cen (ephemeris 2445073.6166 + 1.51997E, Petersen & Hansen 1984); this star is at the centre of the Cepheid II Hertzsprung progression, note the symmetrical maximum.

these stars. For instance Nemec & Lutz (1993) have recently revived Arp's suggestion that not all Type II Cepheids are pulsating in the fundamental mode.

Type II Cepheids are found in both the halo population of our Galaxy (including globular clusters) and in the old-disc population. Their masses are likely to be of order 0.6 $M_\odot$ (see Gingold 1984) and their periods are in the range 0.75 to 40 days. They pulsate radially and studies of their spectra show that shock waves are being propagated outwards through their atmospheres during each pulsation cycle. The shapes of Cepheid II light curves show a

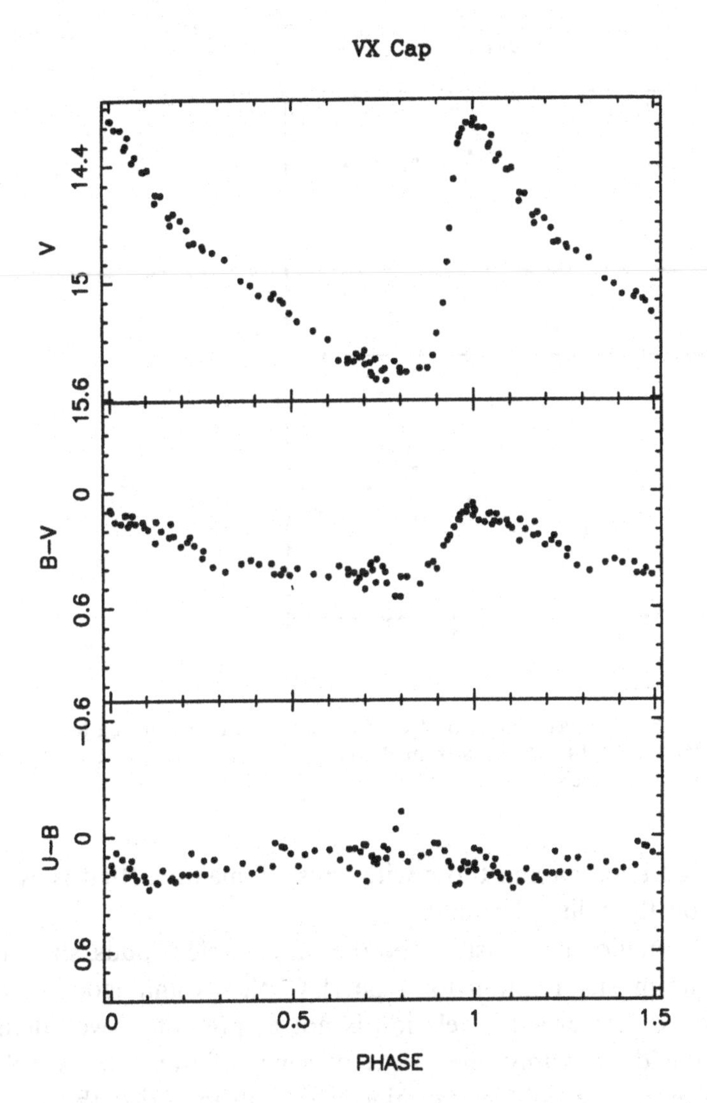

Figure 3.28 $V, B-V, U-B$ phase diagram of VX Cap (Cepheid II; RRd according to Kwee & Braun 1967); ephemeris $2441890.911 + 1.3275497E$. Note the steep rising branch and relatively smooth light curve.

general dependence on period, analogous to the Hertzsprung progression in classical Cepheids discussed in Section 3.7 (Stobie 1973). In the case of the type II Cepheids, the bump in the light curve is generally on the descending branch for periods shorter than about 1.5 days and on the rising branch in longer period stars. However the interpretation of this result is still not entirely clear (see Simon 1986, Petersen & Diethelm 1986, Petersen 1993). Kwee (1967) divides type II Cepheids with periods between 13 and 20 days into two groups:

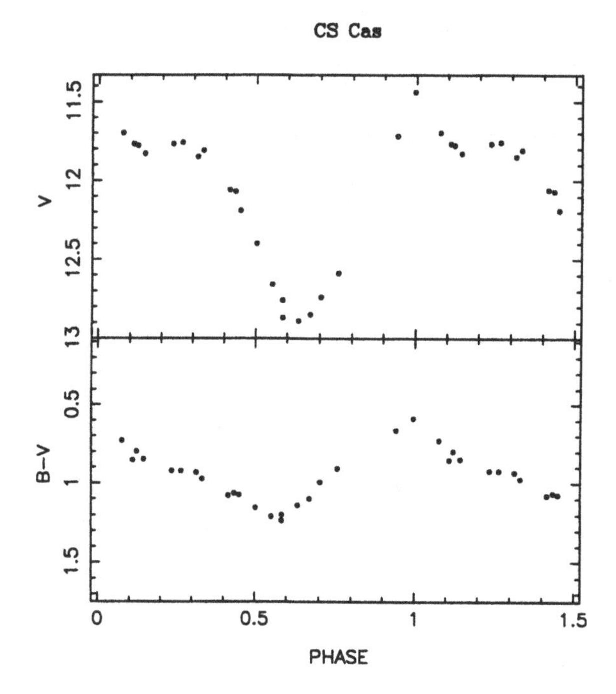

Figure 3.29 $V, B-V$ phase diagram of CS Cas (Cepheid II) according to ephemeris $2436843.80 + 14.74E$ (data from Bahner & Hiltner 1961). Note the bump after the maximum (Kwee's crested form).

those with flat-topped maxima and those with 'crested' maxima (that is, with a shoulder or bump on the falling branch).

Data, especially from globular clusters and the Magellanic Clouds, show that there is a period–luminosity relation for type II Cepheids analogous to that for classical Cepheids. However the relation is not at present as well defined as for classical Cepheids, possibly due to the presence of overtone as well as fundamental pulsators in the samples examined. At a given period the Type II Cepheids are fainter than their classical analogues.

Dwarf spheroidal galaxies in the local group, the Magellanic Clouds and some galactic globular clusters contain 'anomalous Cepheids' which are rather similar to the type II variety but brighter at a given period. These stars are likely to be more massive than Cepheid IIs, perhaps having masses between 1.0 and 1.6 $M_{\odot}$. Their presence in very old systems where only lower mass stars are expected has been explained by postulating their formation in the merger of the two components of a binary star.

The figures show typical type II light curves. The points on several of these curves shown are rather sparse (e.g., CS Cas, Fig. 3.29 and W Vir, Fig. 3.30). The reason for this is that the points plotted in each case are from a single cycle

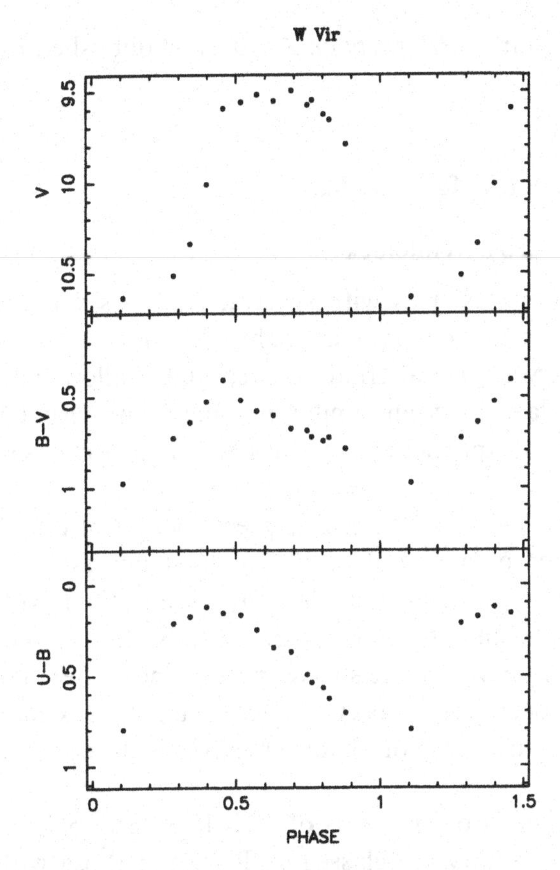

Figure 3.30 $V, B-V, U-B$ phase diagram of W Vir (Cepheid II) according to ephemeris $2430000.00 + 17.2768E$ (Kwee & Braun 1967). Note the shape at maximum (an example of Kwee's flat-topped class).

only. The periods and light curves of these stars vary significantly from cycle to cycle, making a mean over several cycles less informative. In general the light curves of type II Cepheids are less reproduceable than those of classical Cepheids and this itself is sometimes helpful in classifying these stars. One expects in principle that the light curve of a pulsating star will depend on the star's mass, structure and chemical composition. However in practice it may be difficult to unambiguously classify an individual star as a Type II Cepheid, an anomalous Cepheid, or a classical Cepheid, on the basis of the light curve alone. Currently other information (e.g., galactic position, radial velocity, luminosity, chemical composition) is often used together with the light curve to classify stars.

Further useful references are Harris (1985), Nemec (1989), Petersen & Hansen (1984), Petersen & Andreasen (1987), Wallerstein & Cox (1984), and the triannual reports on Variable Stars in *'Reports on Astronomy'* published by the IAU.

3.9 RV Tau variables

P.A. Whitelock

The RV Tau variables are luminous stars with spectral types, as determined from the atomic lines, of F or G at maximum light and G or early K at minimum. The visual light curves have alternating deep and shallow minima, with a period of 30 to 150 days between similar minima. The visual light amplitude is usually between one and two magnitudes although it may exceed three magnitudes.

The phase of the $U-B$ and $B-V$ colour curves precedes that of the visual light curve by up to a quarter of a period. RV Tau light curves are only semi-regular and considerable variation is seen from one cycle to the next for a given star. The longer-period objects – R Sct (Figs. 3.31 and 3.32, P$\sim$144 days) in particular – tend to be less regular than the shorter period ones. Interchanges of deep and shallow minima occasionally occur. These interchanges may be abrupt or gradual. Stretches of irregular or chaotic behaviour have also been recorded.

The RV Tau stars are subdivided on the basis of their long-term behaviour. Those with clear long-term variability are classed as RVb and those without such variations as RVa.[1] The RVb type are periodic with periods in the range hundreds to thousands of days. The RV Tau stars can be oxygen- or carbon-rich and have also been subdivided into groups A, B and C on the basis of their spectra.

Examples of RV Tau stars are found in the globular clusters with particularly low metallicities where they are among the most luminous stars in their respective clusters. It seems likely that the majority of field RV Tau stars do not have as low metallicity as do the cluster stars. In fact, the field RV Tau stars seem to have a range of physical properties and the cluster stars represent one extreme of this range. In a cluster H–R diagram, RV Tau stars fall above the horizontal branch to the high temperature side of the asymptotic giant branch (AGB). There is some suggestion that the field RV Tau stars are on average less luminous than those found in clusters.

The evolutionary status of RV Tau stars is unclear, but they do have very extended atmospheres and are undergoing mass loss. TiO absorption

1. Note that in the *GCVS* the subtypes RVa and RVb are rendered as RVA and RVB. This usage is not recommended, because of possible confusion with the spectroscopic subtypes.

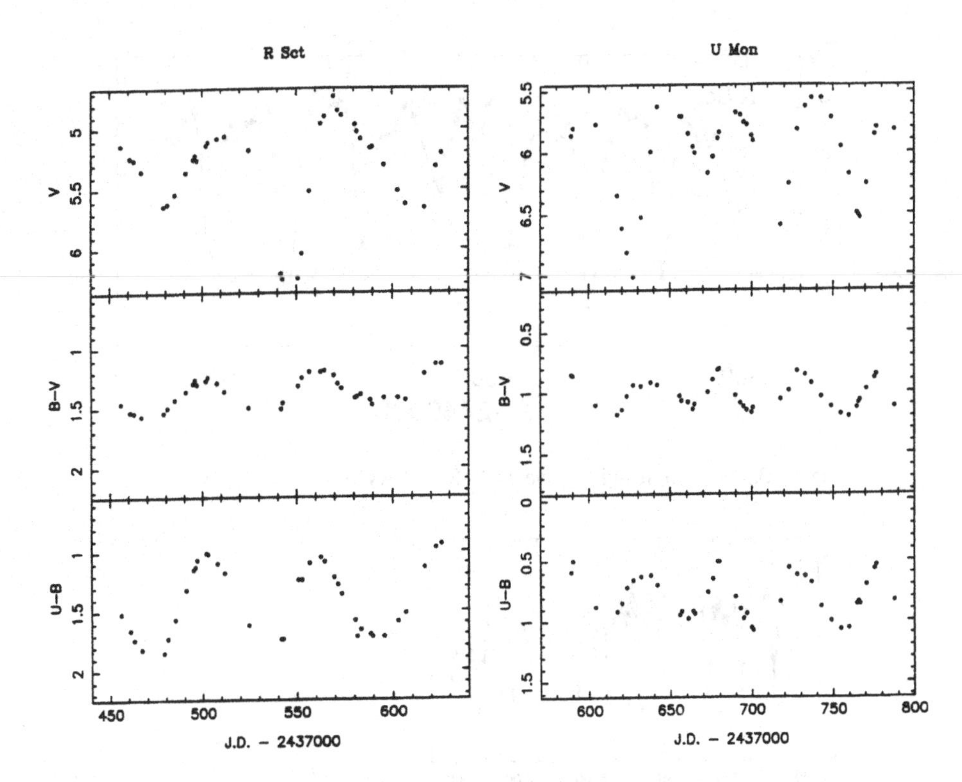

Figure 3.31 $V, B-V, U-B$ light curves of R Sct (RVa), the period is 144^{d} (Preston *et al.* 1963), and U Mon (RVb), the period is $92^{d}_{.}3$ (Preston *et al.* 1963).

bands are sometimes observed in the optical spectra near minimum light. These bands indicate a much later spectral type, perhaps M2 or later, than do the atomic lines and are probably formed high above the photosphere in the extended atmosphere. Some, but not all, RV Tau stars are surrounded by extensive dust-shells as evidenced by their strong infrared emission. They may be AGB stars executing blue loops in the H–R diagram following a helium-shell flash, or they could be post-AGB stars in the process of losing the last remnant of their atmospheres as they turn into white dwarfs.

These variables are closely related to the type II Cepheids (BL Her and W Vir stars) which are also found in metal-deficient globular clusters, and occupy the same instability strip in the H–R diagram at lower luminosity and shorter periods. They also have similarities to the SR variables, in particular the SRd and the UU Her groups.

The illustrative examples are R Sct (RVa, Figs. 3.31 and 3.32) and U Mon (RVb, Fig. 3.33).

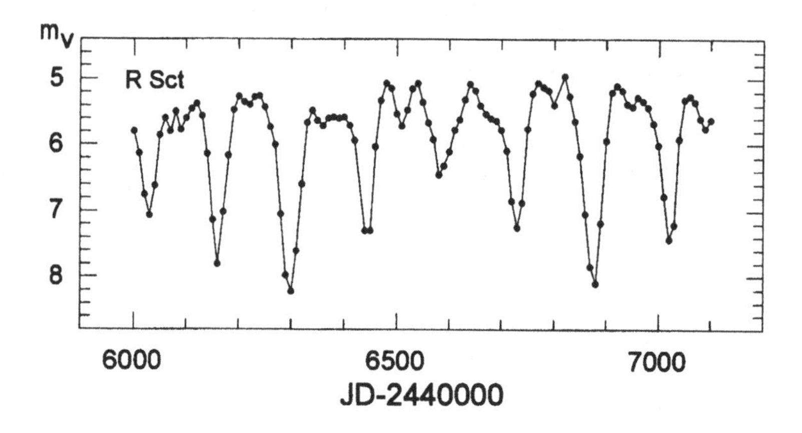

Figure 3.32 Visual light curve of R Sct (AAVSO).

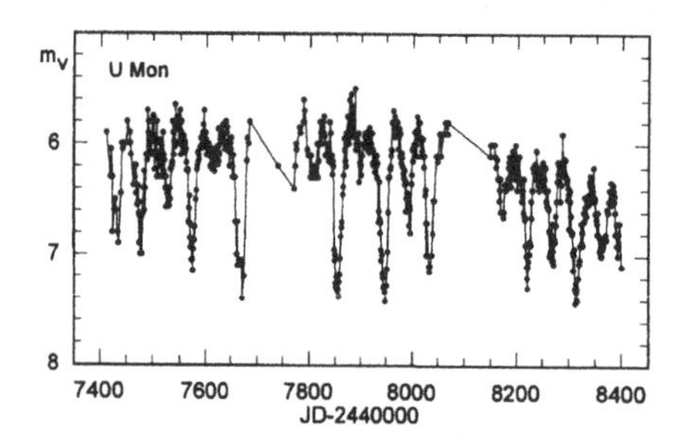

Figure 3.33 Visual light curve of U Mon (RVb). Data from AAVSO.

3.10 Semi-regular and slow irregular variables

P.A. Whitelock

Variability appears to be a fundamental characteristic of cool luminous stars and finding a constant late-type star to use as a photometric standard can present quite a problem. Indeed, it seems likely that almost all of them are variable at some level, though some clearly have larger amplitudes than others. The *GCVS* distinguishes four classes of semi-regular variable: SRa, SRb, SRc and SRd. There is also an SR class for poorly-studied objects. The SRa and SRb stars are giants while the SRc stars are supergiants. The major difference between the SRa class and the Miras is that an SRa may have a visual light amplitude of less than 2.5 magnitudes. In principle, the light curves can also be less regular than those of Miras, but – given that Mira light curves are not necessarily very regular – this is a subjective criterion and may depend on how long the observations of a given star are continued. The SRb class is like the SRa class, but with less obvious periodicity.

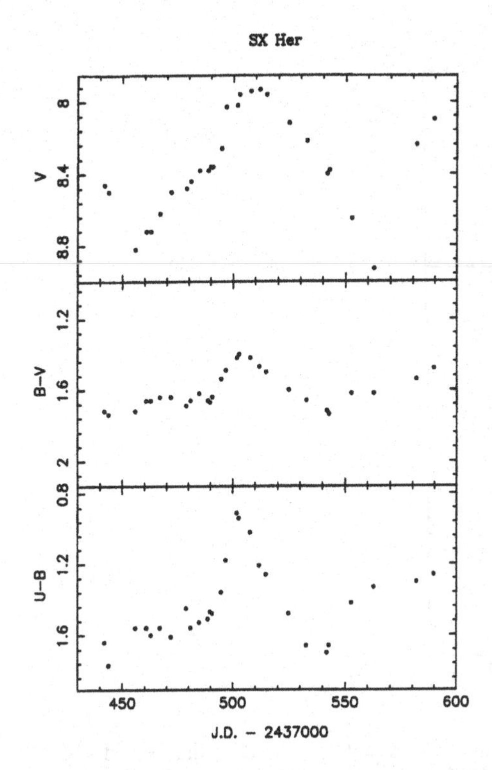

Figure 3.34 $V, B-V, U-B$ light curve of SX Her (SRd), the period is $102^{d}_{.}9$ (Preston *et al.* 1963.)

Variables in classes SRa and SRb include stars of spectral type M, S and C. However, the fraction of carbon semi-regulars is much higher than the fraction of carbon Miras. This is largely a consequence of the way the atmospheric molecular absorption changes with phase. The TiO absorption, which dominates the spectra of the cool oxygen-rich stars, becomes very strong at low temperatures, i.e., at minimum light, and thereby tends to increase the amplitude of the variability allowing such stars to meet the amplitude qualification as Miras. The carbon-rich molecules do not show the same extreme changes, which results in the C-stars falling into the lower amplitude SR category. The SRa and SRb variables may or may not show emission lines; those which do show them tend to be similar to the Miras.

From an evolutionary standpoint the SRa and SRb variables are much more heterogeneous than the Miras and obviously include stars with a range of masses. The globular clusters contain a number of such stars which are on the giant or asymptotic giant branch though always less luminous than the Miras. Kinematic studies of local SRa and SRb variables indicate that, statistically, they are younger than the globular clusters. The SRc stars are

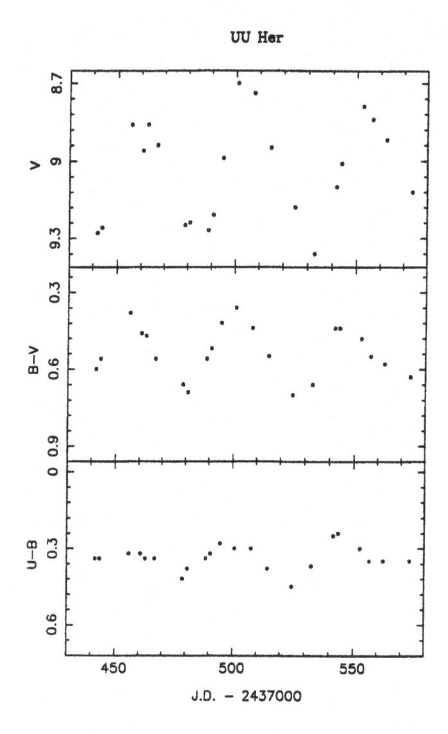

Figure 3.35 $V, B-V, U-B$ light curve of UU Her (SRd), the period is 432ᵈ.7 (*GCVS*, the data are from Eggen 1977).

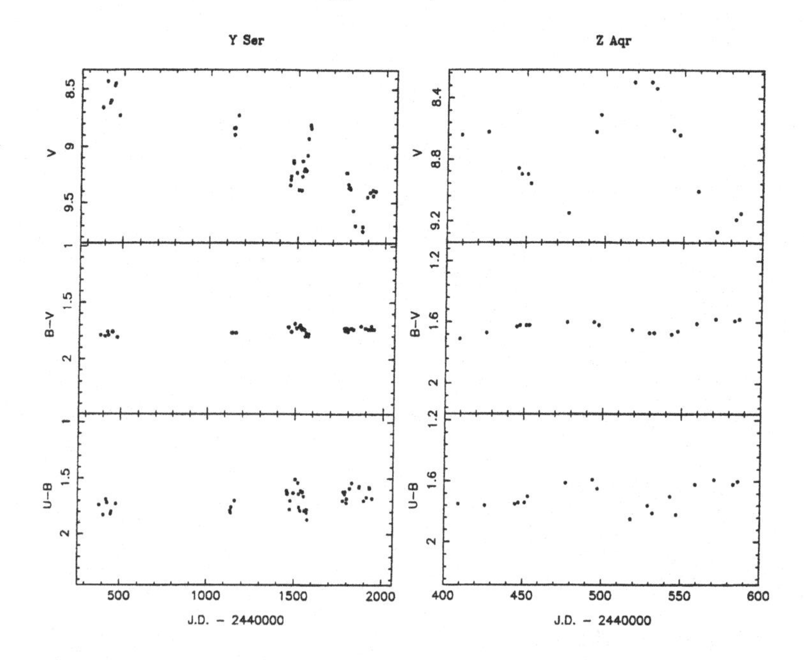

Figure 3.36 $V, B-V, U-B$ light curve of Y Ser and Z Aqr (SRa), the period of Y Ser is 432ᵈ.7 (Ahnert 1950). The data are from Eggen (1977).

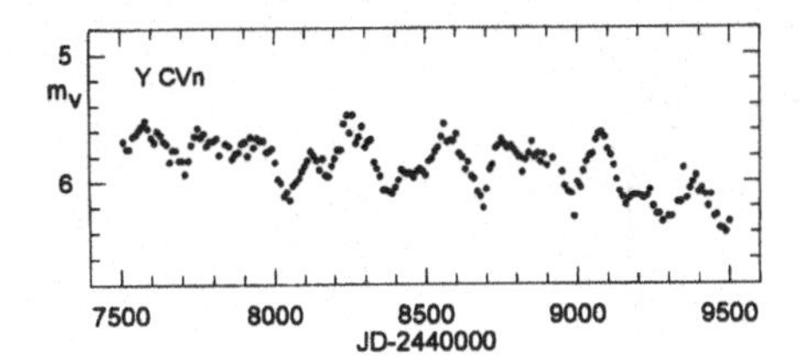

Figure 3.37 Visual light curve of Y CVn (SRb). Data from AAVSO.

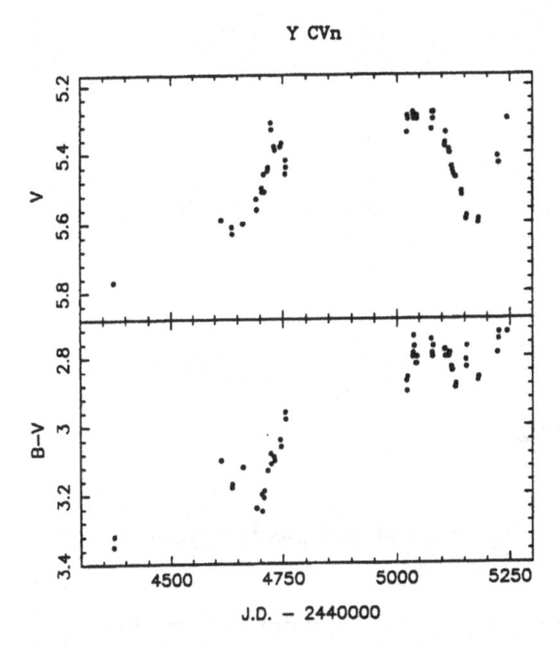

Figure 3.38 $V, B - V$ light curve of Y CVn (SRb), Vetesnik (1984). The main period is 157 days (Gaposchkin 1952).

generally thought to be massive with progenitors in excess of about 8 $M_\odot$. They follow an approximate period–luminosity relation.

The SRd stars are semi-regular giants and supergiants of spectral type F, G or K, sometimes with emission lines. Although a small group, they are again heterogeneous and in general poorly studied. Those with emission lines and large amplitude variations have been suggested as a low-metallicity analogue of the Mira variables. Such stars are found in globular clusters. Others seem more closely related to RV Tau stars. A subgroup, sometimes called UU Her stars, is found at high galactic latitude and has been proposed as a possible transition phase between the asymptotic giant branch and the white dwarf evolutionary phases. Alternative explanations involve a close binary with a common envelope.

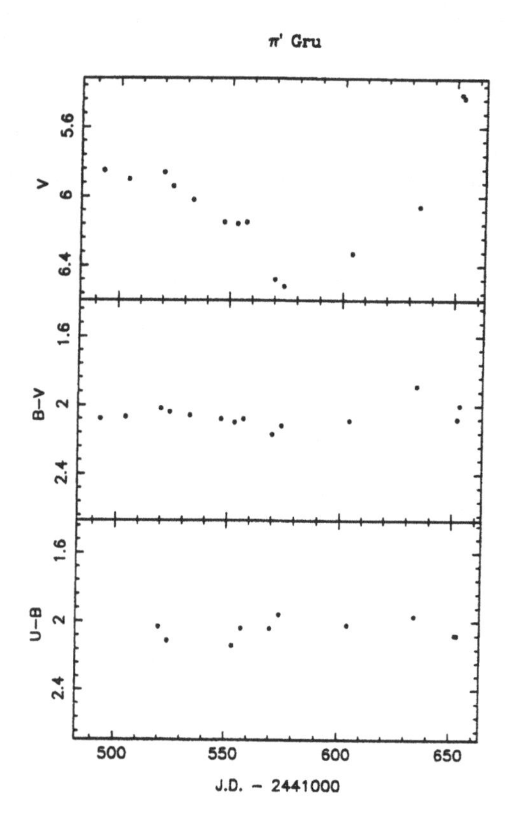

Figure 3.39 $V, B-V, U-B$ light curve of π^1 Gru (SRb) (Eggen 1975).

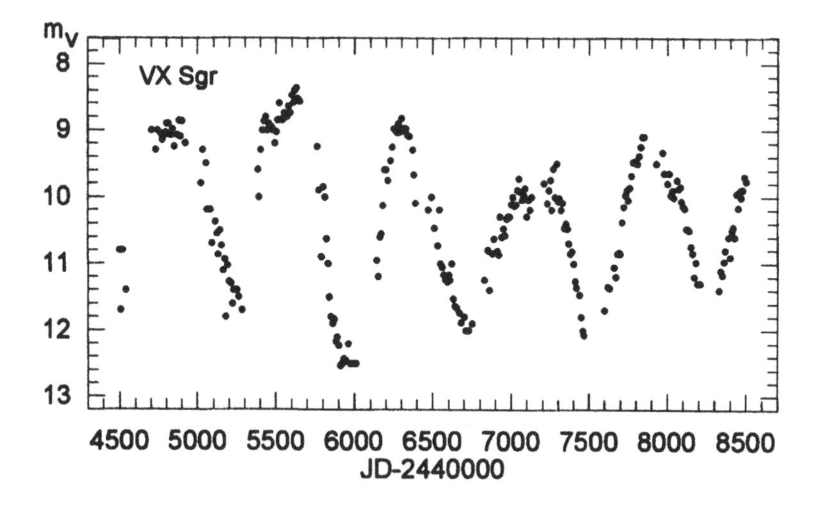

Figure 3.40 Visual light curve of VX Sgr (SRc), AAVSO.

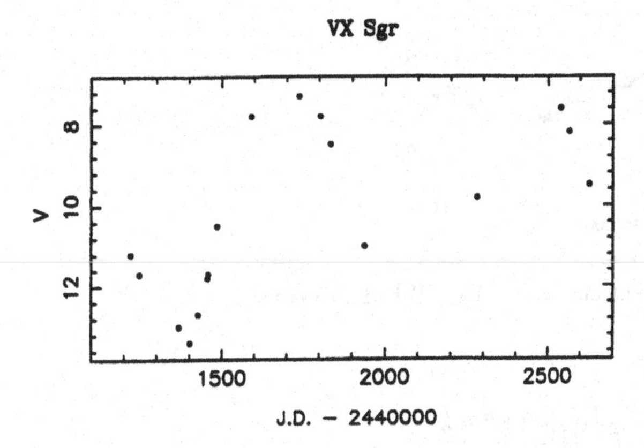

Figure 3.41 V light curve of VX Sgr (SRc), the period is 732 days. Kukarkin *et al.* (1955), data are from Lockwood & Wing (1982).

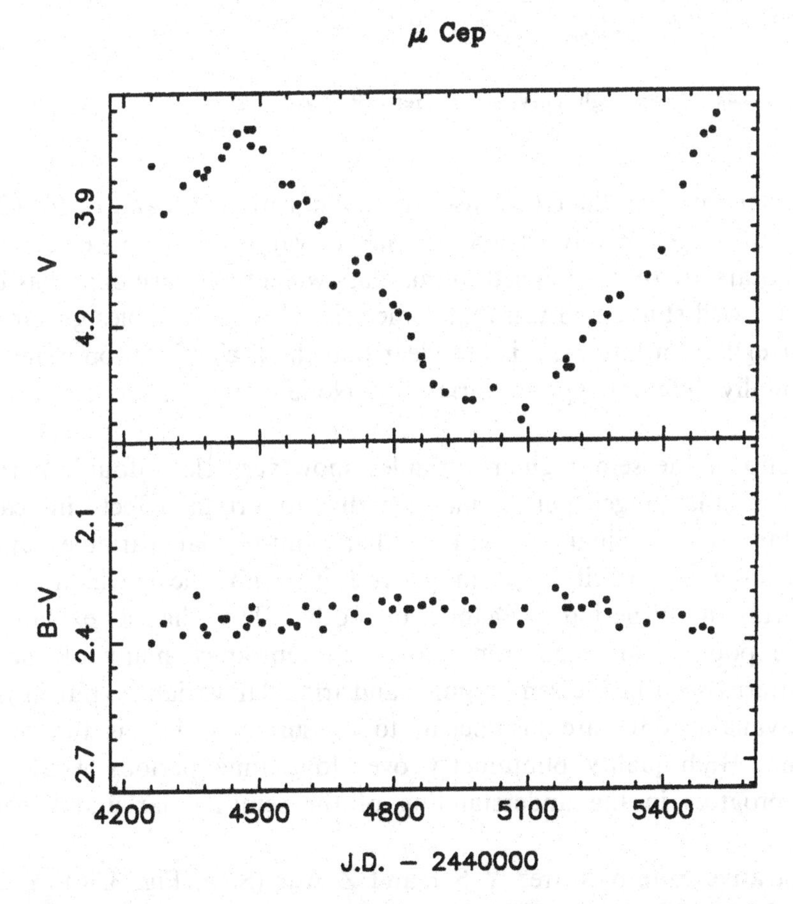

Figure 3.42 $V, B - V$ light curve of μ Cep (SRc), the period is 730 days (Polyakova 1975, the data are from Polyakova 1984).

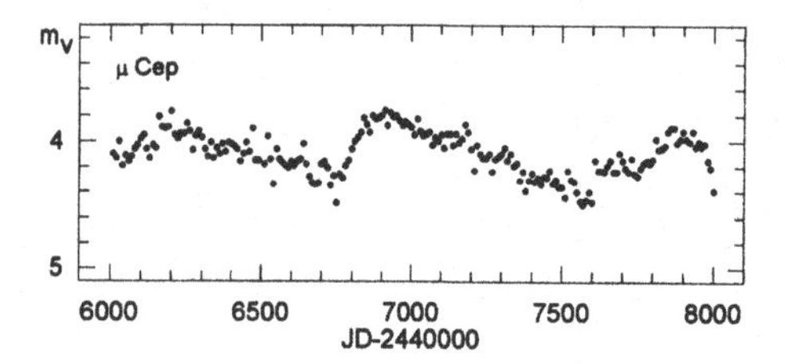

Figure 3.43 Visual light curve of μ Cep, (SRc), AAVSO.

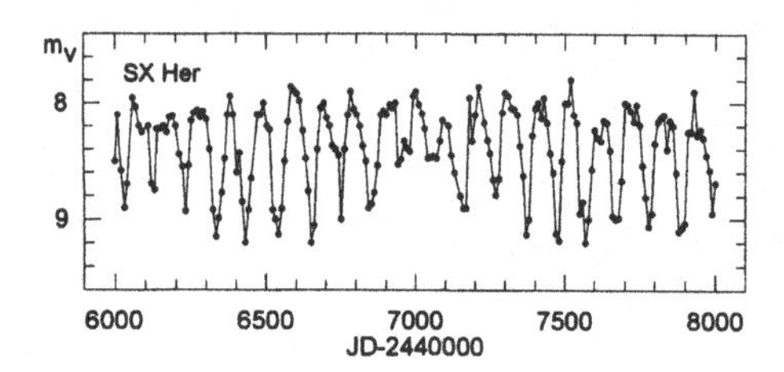

Figure 3.44 Visual light curve of SX Her (SRd), AAVSO.

For irregular variables the *GCVS* uses the classifications Lb and Lc for giants and supergiants, respectively. These are slowly varying with no evidence of periodicity. Stars are often assigned to this class when their variability has been noted but not well-studied. Given that some semi-regular variables go through phases of irregular variations, it is not clear that the L classification represents a fundamentally different type of variability. None of these stars is illustrated here.

A subgroup of the semi-regular variables show very clear double periods. In some cases the longer period may be due to orbital effects indicating that the star is in a binary system. Other semi-regular variables apparently show multi-periodicity, but in general it is not clear whether these stars are truly multi-periodic, chaotic or both. The chaotic explanation has become popular in recent times and can reproduce many of the behaviour patterns seen in the semi-regular and irregular variables, but in most cases the available data are insufficient to be sure that this is the correct explanation. High-quality photometry over long time periods would certainly aid progress in the understanding of the semi-regular and irregular variables.

The illustrative examples are: Y Ser and Z Aqr (SRa, Fig. 3.36), Y CVn (Figs. 3.37, 3.38) and π^1 Gru (SRb, Fig. 3.39), VX Sgr (Figs. 3.40, 3.41) and μ Cep (SRc, Figs. 3.42, 3.43), and SX Her (Figs. 3.44, 3.34) and UU Her (SRd, Fig. 3.35). For more detailed reading, we refer to Johnson & Querci (1986).

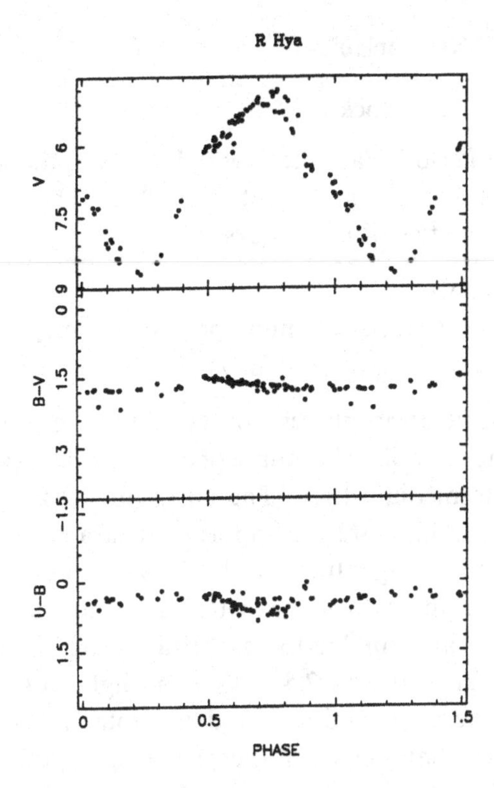

Figure 3.45 $V, B-V, U-B$ light curve of R Hya (Mira), the period is 388^{d}.9 (*GCVS*). The data are from Eggen (1975), Celis (1977) and Nakagiri & Yamashita (1979).

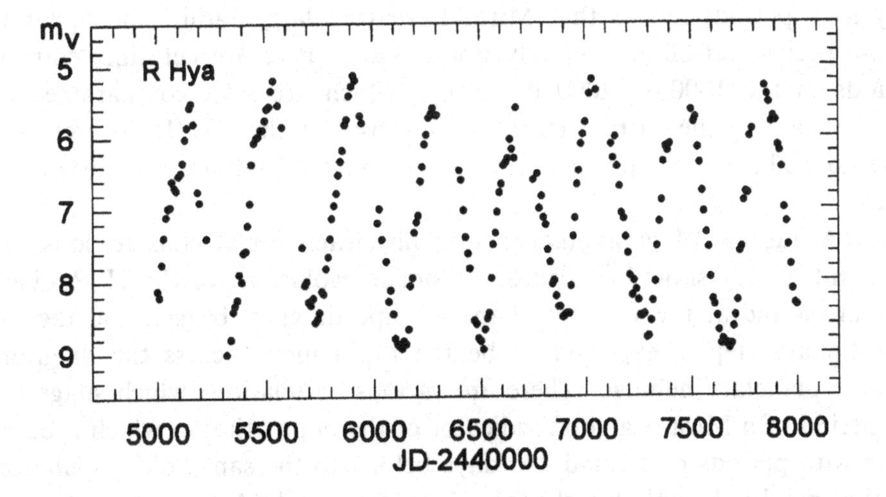

Figure 3.46 Visual light curve of R Hya, AAVSO.

3.11 Mira variables

P.A. Whitelock

The Miras (also known as Long-Period Variables, or LPVs) are the most homogeneous and best studied of the pulsating red variables. The *GCVS* distinguishes three defining characteristics of these stars:

(i) the spectral type is Me, Se or Ce
(ii) the visual or photographic light amplitude must be over 2.5 mag
(iii) the period should be in the range 80 to 1000 days.

The spectral type tells us that Mira atmospheres contain strong molecular absorbtion features and are therefore cool. The atmosphere may be oxygen-rich (Me), carbon-rich (Ce) or intermediate (Se). The emission lines, whose presence is signified by the 'e' in Me etc., are an important characteristic of this type of variability as they are the signature of shock waves associated with pulsation. The amplitude cut-off is to some extent arbitrary and results in a few stars which are physically similar to the Miras being classified as SRa because their amplitudes fall short of 2.5 mag. The light curves in the infrared (where most of the energy is emitted) and the total integrated luminosity have smaller amplitudes than those of the visual light, although they are mostly over 0.5 mag. The large visual amplitudes thus arise from the combination of the fact that we are observing temperature variations from the blue side of the star's energy-distribution peak and from changes in molecular-band strengths associated with these temperature changes. The very long periods tell us that Miras have very large radii. The upper limit to the period cut-off is probably not useful. There are certainly stars with periods in the 1000 to 2000 day range which are in a comparable evolutionary state to the Miras; these are a subset of the OH/IR sources which have no visible counterpart as they are surrounded by thick circumstellar dust shells.

The Miras are of great interest astrophysically for various reasons. They represent a very short-lived phase in stellar evolution; in the H–R diagram they are found at the very tip of the asymptotic giant branch and their next evolutionary step is expected to be the rapid move across the diagram to become planetary nebulae. There are various correlations which suggest that the period of a Mira is a good indicator of the population to which it belongs. Stars with periods of around 200 days belong to the same, old population as do the metal-rich globular clusters. Longer period Miras are more massive and/or more metal rich. Consistent with this picture – but contrary to popular belief – there is no evidence that Miras systematically evolve to longer periods

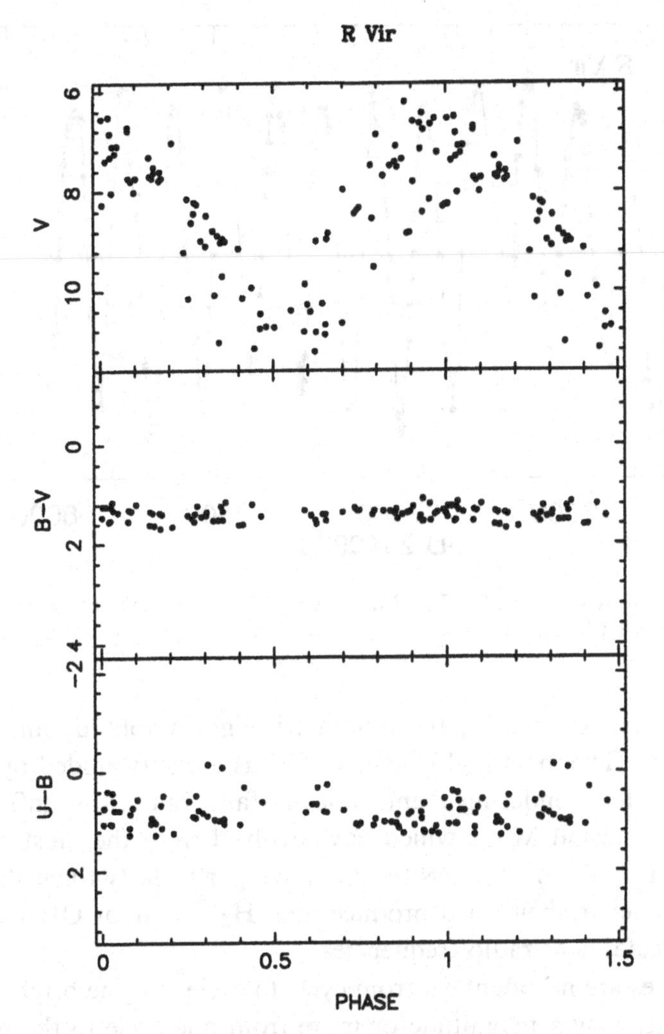

Figure 3.47 $V, B-V, U-B$ phase diagram of R Vir (Mira), the period is 145^{d}.6 (*GCVS*). The data are from Eggen (1975), Celis (1977) and Barnes (1973).

as they age. Miras are also useful as distance indicators as they obey a period–luminosity relation, which can be expressed either in terms of the total (i.e. bolometric) luminosity or in terms of the near-infrared magnitude (usually K, i.e. 2.2 μm).

It remains uncertain whether Miras pulsate in the fundamental or first overtone modes. While there are theoretical reasons for favouring the fundamental mode, the observational evidence favours the overtone. Miras are losing mass rapidly (10^{-8}–10^{-4} $M_{\odot}$ yr^{-1}), although the mechanism for this, as for the pulsation itself, is not well understood. The mass-loss rates are statistically

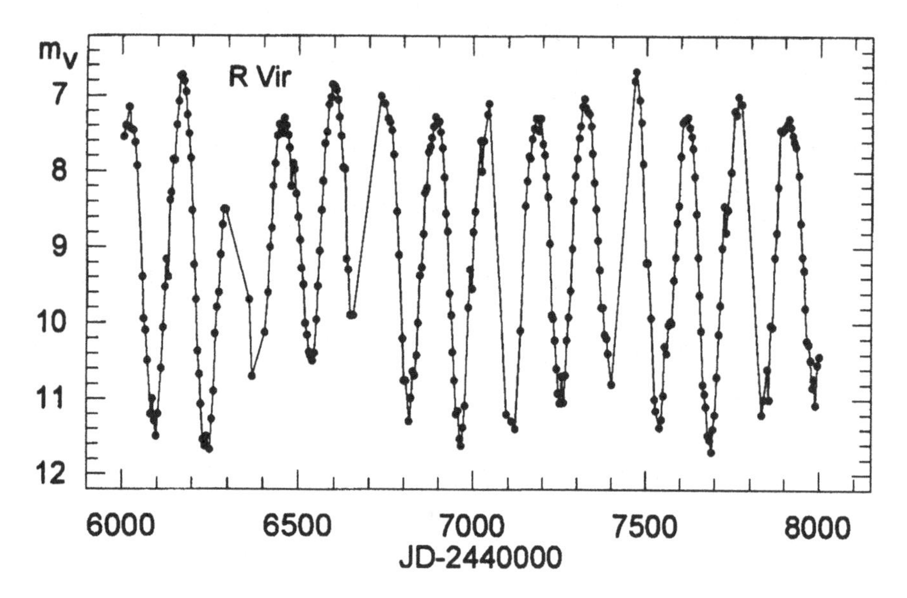

Figure 3.48 Visual light curve of R Vir (Mira), AAVSO. For the sake of clarity, the dots are connected by a line.

correlated with the pulsation period, the bolometric light amplitude and the shape of the light curve. The most highly evolved Miras are surrounded by the material they have ejected, rendering them optically faint but strong infrared sources. The very long-period Miras which have evolved from the most massive progenitors and have the most mass to lose, have particularly thick shells. Some of these circumstellar shells also produce SiO, H_2O, and/or OH maser emission which is detectable at radio frequencies.

The Mira light curves are not identical from cycle to cycle and the brightness at maximum often varies by a magnitude or more from one cycle to the next. Period changes are observed in certain stars and are seen particularly clearly in R Aql and R Hya, which may be undergoing helium shell flashes.

The illustrative examples are R Hya (Figs. 3.45 and 3.46) and R Vir (Figs. 3.47 and 3.48). For further reading, we refer to Campbell (1955), Johnson & Querci (1986), Willson (1986) and Whitelock *et al.* (1994).

3.12 ZZ Ceti variables

H.W. Duerbeck

ZZ Ceti stars are non-radially pulsating white dwarfs, which change their brightness with periods between 30 seconds and 25 minutes. The amplitudes can reach $0^{\mathrm{m}}.2$ in V. The *GCVS* gives three subtypes, classified according to spectral type:

(i) ZZA – spectral type DA: white dwarfs with hydrogen absorption lines (e.g. ZZ Cet)
(ii) ZZB – spectral type DB: white dwarfs with helium absorption lines (e.g. V 777 Her)
(iii) ZZO – spectral type DO: white dwarfs with continuous spectra or PNNV (variable planetary nebula nuclei) (e.g. GW Vir).

The ZZ Ceti stars cannot be radial pulsators, since their periods are too long. Multicolour observations have confirmed that the pulsation modes are nonradial gravity (=g) modes. Contrary to the pressure (=p) modes, the g-modes are produced by fairly horizontal motions. Several periods are simultaneously excited, their frequencies are often split into close pairs by the slow rotation of the star.

The periods can be extremely stable ($\Delta P/P \leq 10^{-12}$); unstable periods with period changes occuring over a few hours are probably caused by interactions of various periods (beating of closely-spaced frequencies).

The single white dwarfs have masses near 0.6 $M_{\odot}$, they have CO-nuclei and thin H (DA) or He (DB) layers. The variable white dwarfs of types ZZB and ZZA have partial ionisation zones of He and H in their surface layers, the white dwarfs of ZZO type have partial ionisation zones of C and O. It is possible that the κ-mechanism, which is also responsible for the Cepheid pulsations (see also Section 3.2), operates in these layers. Pulsations of a thin CO-convection zone have also been suggested for the behaviour of all three groups (Cox 1993).

Table 3.1 Observed properties of pulsating white dwarf stars. A is the fractional amplitude. In the last two columns the numbers in brackets indicate the range of values, a single number indicates a typical value

Class	Spectra	$\log g$	$\log L/L_{\odot}$	T (K)	P (s)	A
PNNV	He II,C IV	>6	3–4	> 100 000	(>1000)	
	nebula				1500	0.01
DOV	He II, C IV, O VI[a]	7	2	> 100 000	(300–850)	(<0.001–0.04)
					500	0.01
DBV	He I[b]	8	-1.2	25 000	(140–1000)	(<0.001–0.04)
					500	0.02
DAV	H[b]	8	-2.8	12 000	(100–1200)	(< 0.001–0.1)

[a] Denotes absorption lines with narrow emission core.
[b] Are pure absorption lines.

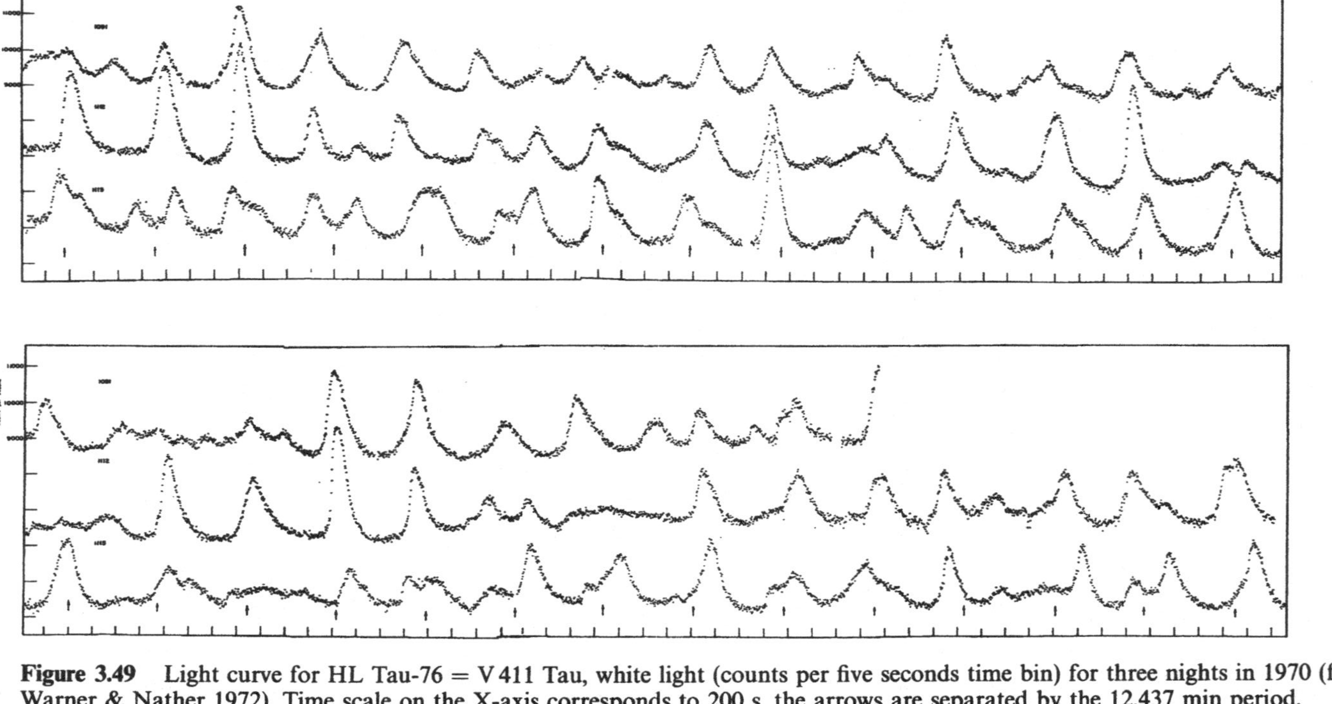

Figure 3.49 Light curve for HL Tau-76 = V 411 Tau, white light (counts per five seconds time bin) for three nights in 1970 (from Warner & Nather 1972). Time scale on the X-axis corresponds to 200 s, the arrows are separated by the 12.437 min period.

Table 3.1, taken from Winget (1988), summarizes the observed properties of the ZZ stars.

The first pulsating white dwarf was discovered by Landolt (1986), when making photometric observations of objects from a list of suspected white dwarfs (Haro-Luyten HL Tau-76 = V411 Tau, see Fig. 3.49)

The *GCVS* lists 22 objects. Five of them occur in nova-like or dwarf nova systems. Of the remaining 17, the majority belong to the ZZA type.

4
Rotating variables

4.1 Ap and roAp stars

C. Sterken

The existence of stars whose surface is severely depleted in He with, at the same time, overabundance of Fe, Si and Cr in spots, has been known since the early days of spectral classification, when the phenomenon was first detected in Ap stars (for details, see Jaschek & Jaschek 1987, Morgan 1933).

Chemically-peculiar (CP) stars, in general, are stars of spectral type B2 – F of which the spectra reveal signatures of chemical peculiarities such as, for example, strongly-enhanced spectral lines of Fe and rare-earth elements. In this group, there is a *magnetic sequence* – referring to, as Hensberge (1994) puts it, 'those stars that show a magnetic field that is strong and global (a large dipolar contribution to the field), so that it is detectable with the present [observing] techniques. It does not imply that HgMn stars, or metallic-line (Am) stars, etc. would have no magnetic field at all. Stars in the non-magnetic sequence may be either without field, with a significantly weaker global field, or with a strong field of complicated structure, such that the measurable effect, averaged-out over the visible disc, is insignificant'. Ap stars have global surface magnetic fields of the order of 0.3 to 30 kG (thousands of times stronger than that of the sun), and the effective magnetic-field strength varies with rotation, a situation that led to interpretation in terms of the oblique-rotator model in which the magnetic axis is oblique to the rotation axis (this model was first suggested by Stibbs in 1950). The time scales of light variations seen in Ap stars range from minutes to decades.

Ap stars are intrinsically slow rotators (but the hotter stars rotate faster than the cooler ones), the length of the rotation period can be derived as the aspect of their spotted surface varies periodically with time. Most periods are of the order of one day to one week, with a tail towards longer periods (Hensberge

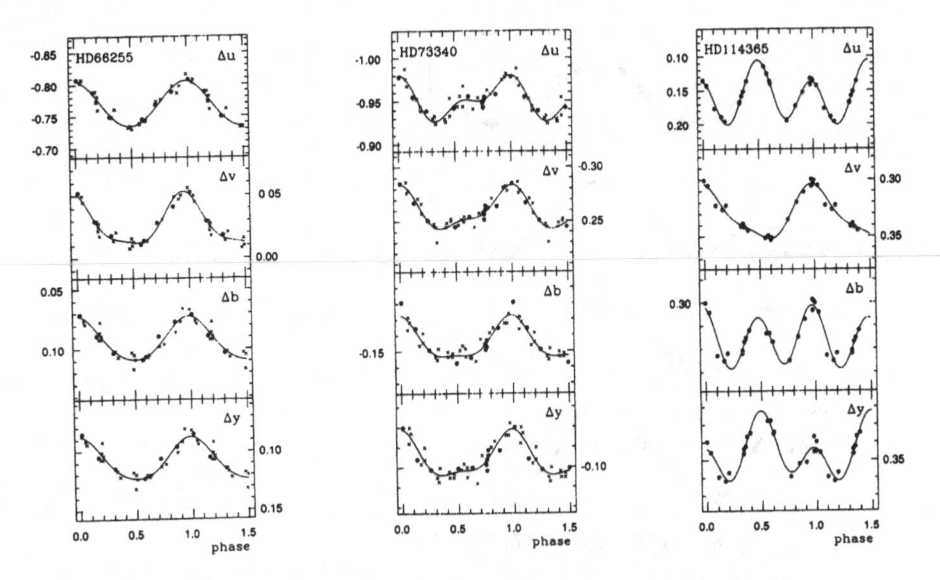

Figure 4.1 Differential *uvby* phase diagrams of HD 66255, HD 73340 and HD 114365 (Ap) for ephemeris JD_{uvby} max = 2448343.439 + 6.8178*E* (left), JD_{uvby} max = 2448345.545 + 2.667588*E* (middle) and JD_v max = 2448347.832 + 1.271925*E* (right). The continuous line is a least-square fit (Catalano & Leone 1991).

1994, see also Fig. 4.1). Other sources of variability, such as binary motion or pulsation (roAp stars, see below) may be superposed.

Periods of the order of 100 days can be associated with very slow rotation, but for periods of the order of years to tens of years the explanation in terms of rotation is less well-accepted. Surface inhomogeneities and the magnetic structures on Ap stars seem to be quite stable over the years. Figure 4.2 illustrates the interesting case of TW Col where the phase diagram shows a small-scale dip that is in conflict with the spot-rotation model.

Catalano *et al.* (1991) discovered CP stars to be variable also in the IR, the near IR light curves seem to be phase related to the magnetic field variations in the sense that magnetic-field extrema might coincide in time with the IR light extrema.

A very interesting system is AO Vel, a southern multiple system consisting of three probably fairly identical stars, two of which are from a detached eclipsing binary while the third is in a wide orbit. The catalogue of Catalano & Renson (1984) lists AO Vel as being one of the very few (< 5) known Ap eclipsing binaries, and among those it is the only example of the Si-type. A photometric analysis and orbital parameters are given by Clausen *et al.* (1995), see Fig. 4.3 for magnitude and colour light curves.

The rapidly-oscillating Ap stars are cool magnetic Ap SrCrEu stars which

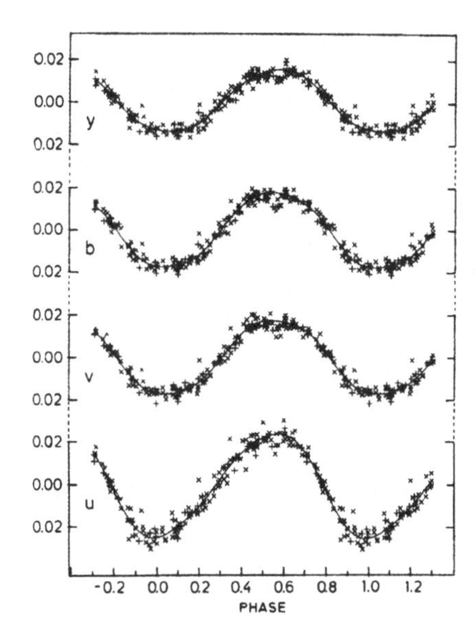

Figure 4.2 Differential *uvby* phase diagrams of TW Col (Ap), ephemeris 2444600.7 + 1.37853E (Renson & Manfroid 1992). The curve is a four-frequency sine fit to the observations outside the phase interval 0.5 – 0.59 during which a dip occurs.

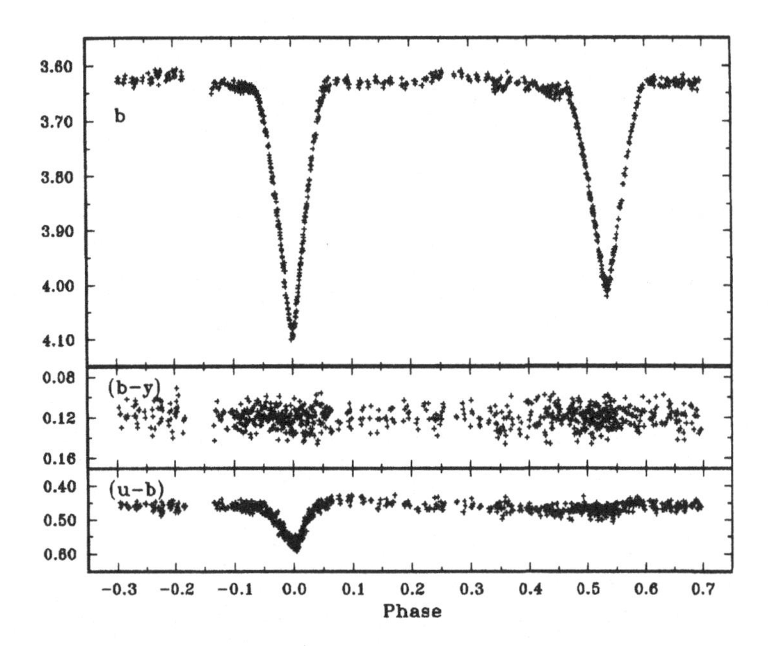

Figure 4.3 $b, b-y, u-b$ phase diagrams of AO Vel, $P_{\rm orb} = 25^{\rm y}.6$ (Clausen *et al.* 1995). The position of the secondary minimum points to an eccentric orbit, the *uvby* light curves are asymmetric around mid-eclipse, an effect attributed to surface inhomogenities (spots).

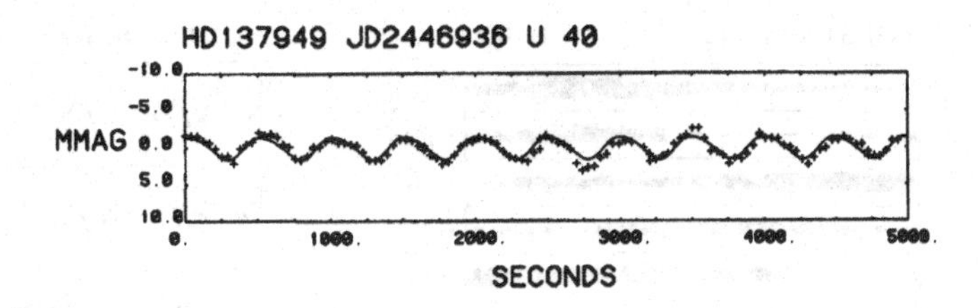

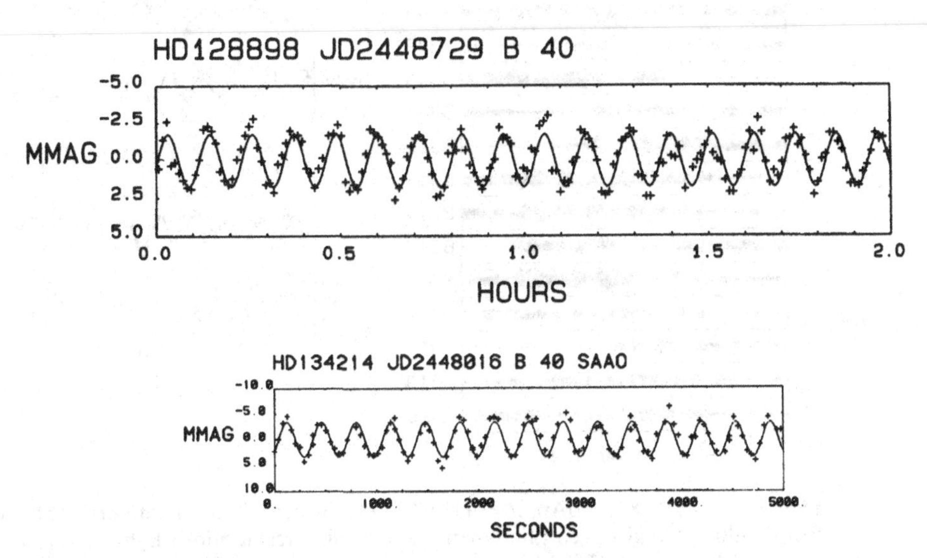

Figure 4.4 Some roAp light curves: Top: 33 Lib, $P = 8^{\rm m}\!.272$ (Kurtz 1991). Middle: a 2-hour portion of the *B* light curve of α Cir, $P_1 = 6^{\rm m}\!.8$ (Kurtz *et al.* 1993a). Bottom: HD 134214, $P = 5^{\rm m}\!.65$ (Kurtz *et al.* 1991). The solid curves are best-frequency fits; note the extremely small amplitudes of the variations (several mmag peak-to-peak).

pulsate in high-overtone, low-degree non-radial modes with periods which range from about 5 minutes to less than 20 minutes. They were first detected by Kurtz (1982). The amplitudes of the light variations are less than a few millimag. These pulsations are dominated by the strong global magnetic fields of the Ap stars: their amplitude is modulated with rotation in phase with the magnetic-field modulation, and this can be described by two models, viz. the oblique-pulsator model and the spotted-pulsator model. The oblique-pulsator model (proposed by Kurtz in 1982) assumes that the pulsation and magnetic axes are aligned and are oblique to the rotation axis; the rotational modulation of the light variations is caused by the varying aspect of non-radial pulsation modes (see Kurtz 1990, Shibahashi & Takata 1993, Kurtz & Martinez 1994). The spotted-pulsator model assumes that the pulsation axis coincides with the rotation axis of the star so that the pulsation modes are always seen from the same viewing angle, but that the ratio of flux to radius variations and the phase

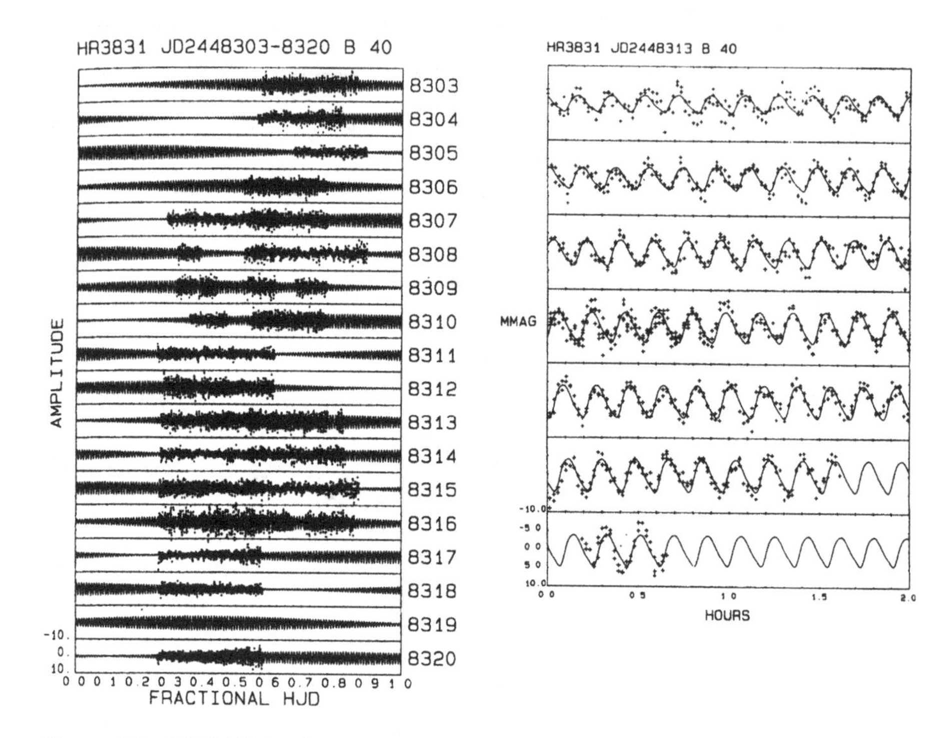

Figure 4.5 HR 3831 (roAp, $P = 11^{\mathrm{m}}.67$). Left: low-resolution light curve obtained from multi-site high-speed photometry. Right: high-resolution B light curve, each panel is 2^{h} long and $0^{\mathrm{m}}.002$ high, the total represents a nearly continuous $12^{\mathrm{h}}.6$ long light curve. The curve is the best-fitting frequency solution (Kurtz et al. 1991).

lag between the flux and radius variations are variable over the surface as a function of the magnetic-field strength (Mathys 1985) leading to the observed amplitude modulation. Less than 30 roAp stars are known, they are all located in the lower δ Scuti-star instability strip (see also Section 3.5).

A very interesting newly-discovered characteristic of the roAp stars is that they have intrinsic, cyclic frequency-variability on a time scale of hundreds of days to years (Kurtz & Martinez 1994). This suggests that a magnetic cycle may be operating in these stars. Such a variation of the pulsation period has also been detected in the ZZ Ceti star G29-38 (see Section 3.12), where one pulsation frequency varies cyclically with a period of 110 days (Winget *et al.* 1990).

For a general review of the A stars, Ap stars and the oblique-rotator model, we refer to Wolff (1983), and to Landstreet (1992) for a general discussion of the magnetic fields in all stars. For recent reviews of the roAp stars, see Shibahashi (1987), Kurtz (1990) and Matthews (1991).

Figure 4.4 shows a selection of roAp light curves: 33 Lib with $P = 8^{\mathrm{m}}.272$

(Kurtz 1991), α Cir (a binary) with $P_1 = 6\overset{m}{.}825$ and $P_2 = 6\overset{m}{.}832$ (Kurtz *et al.* 1993b), and HD 134214 with $P = 5\overset{m}{.}65$ (Kurtz *et al.* 1991). Figure 4.5 shows the entire amplitude modulation repertory of HR 3831, the best-studied roAp star ($P = 11\overset{m}{.}67$), in a low time-resolution graph (Kurtz *et al.* 1991).

4.2 Ellipsoidal variables

D. S. Hall

The ellipsoidal variables are a subgroup of the rotating variable stars. They are non-eclipsing binaries in which neither, one, or both stars is elongated by the mutual tidal forces. As the binary orbits, the elongated star rotates, producing two maxima and two minima per orbital cycle. Most eclipsing binaries show additional variability by this same mechanism, but the *GCVS* does not add the 'ELL' classification to them. Stronger limb-darkening effects on the pointed end of the more elongated star can make one minimum significantly deeper than the other (Hall 1990b).

One could imagine another subgroup of rotating variable stars which would also include non-eclipsing binaries: systems in which the reflection effect dramatically illuminates one hemisphere of one of the two stars. An example would be HZ Her = Her X-1, in which case the X-ray-emitting neutron star brightens the facing hemisphere of its companion (a star of spectral type late

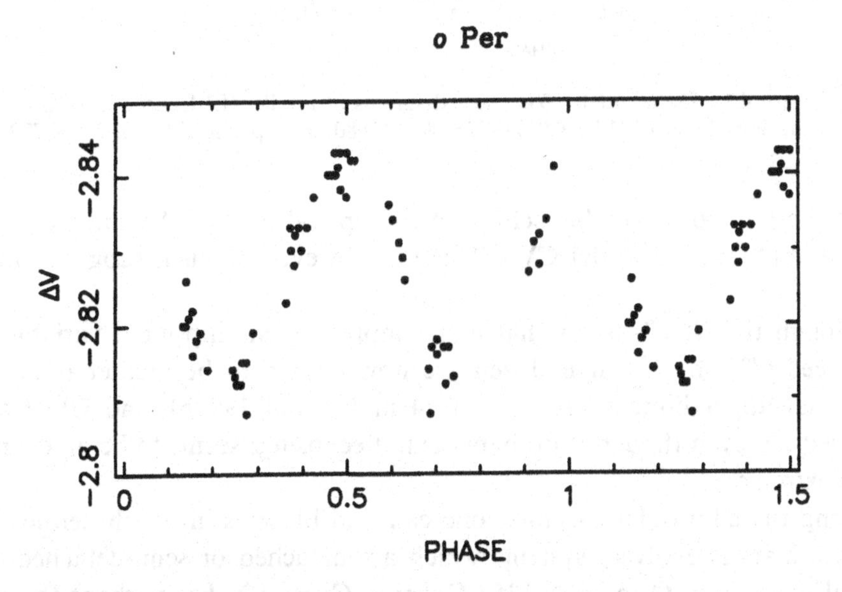

Figure 4.6 Differential *V* phase diagrams of *o* Per (ELL) according to ephemeris 2418217.924 + 4.41916*E* (Lynds 1960). Comparison star: HD 23478

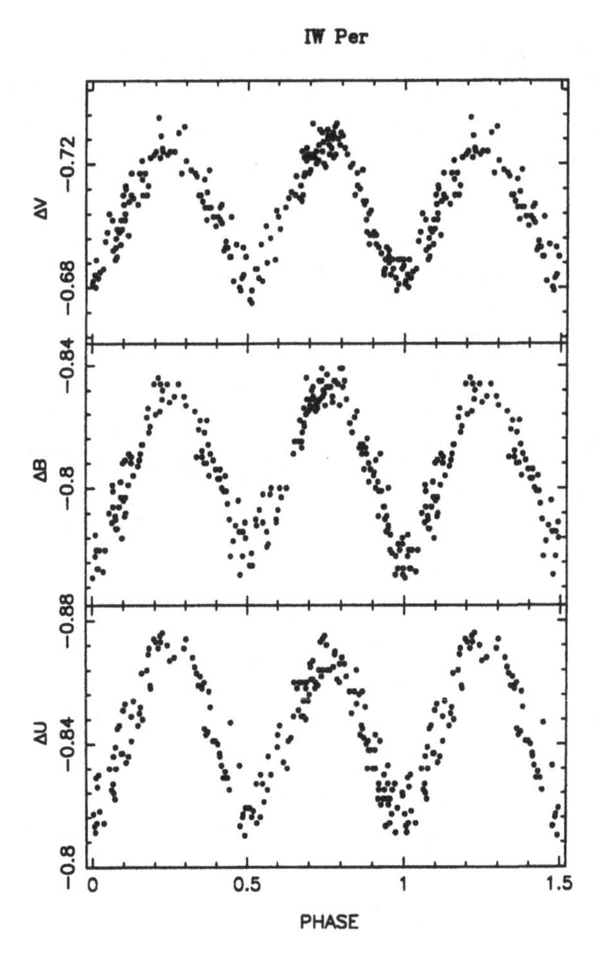

Figure 4.7 Differential *UBV* phase diagrams of IW Per (ELL) according to ephemeris 2433617.317 + 0.9171877*E* (Kim 1980). Comparison star: BD +40 772.

A) but there is no detectable eclipse in the optical part of the spectrum. Another example would be BH CVn. There is, however, no such subgroup in the *GCVS*.

Although the *GCVS* states that light amplitudes in ellipsoidal variables do not exceed $0\overset{m}{.}1$ in *V*, a half dozen are now known to be greater than that. The actual upper limit is more like $0\overset{m}{.}4$ in *V* (Hall 1990b), and UU Cancri (classified EB even though it probably is not eclipsing) seems to be an example of this extreme.

Among the ellipsoidal variables one can find binaries in which neither one, nor both stars is evolved, systems which are detached or semi-detached, and spectral types from O to M. V 1357 Cygni = Cygnus X-1 is perhaps the most famous binary in this group and the recurrent nova T Coronae Borealis varies

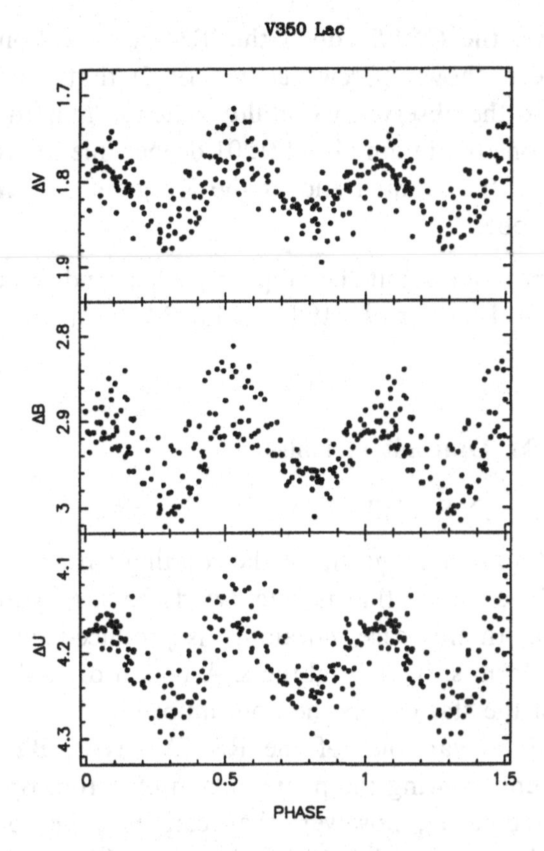

Figure 4.8 Differential *UBV* phase diagrams of V 350 Lac (ELL) according to ephemeris 2446599.82 + 17.755*E* (Boyd *et al.* 1990). Comparison star: HR 8541.

by the ellipticity effect during years between its outbursts, though the *GCVS* classifies only V 1357 Cyg as ELL.

The first ellipsoidal variable discovered, and the prototype of the group, was b Persei. Stebbins (1923) detected the $0^{\mathrm{m}}\!.06$ amplitude during his 1920–1922 photoelectric photometry.

The binaries selected to illustrate variables of the ELL type are

(i) *o* Persei: a B1 III + B2 III binary with an orbital period of $4^{\mathrm{d}}\!.42$. The full amplitude of the ellipticity effect, to which both stars contribute, is $0^{\mathrm{d}}\!.07$ in *V*. The two minima are equal in depth (Fig. 4.6).

(ii) IW Persei: the spectral type is A5m and the orbital period is $0^{\mathrm{d}}\!.92$. The full amplitude of the ellipticity effect is $0^{\mathrm{m}}\!.05$ in *V* and the two minima are equally deep (Fig. 4.7).

(iii) V 350 Lacertae: this single-lined binary has a spectral type of K2 III and an orbital period of $17^{\mathrm{d}}\!.755$ (Fig. 4.8). Because the K2 giant is chromospherically active, the *GCVS* added the 'RS' classification but only recently has it been shown (Crews *et al.* 1995) that starspots contribute appreciably to the observed variability. The full amplitude is about $0^{m}.075$ in V with one minimum about $0^{m}.01$ deeper due to greater limb-darkening effects on the pointed end of the K2 giant, which fills about 3/4 of its Roche lobe.

The interested reader can learn more about the ellipticity effect and the ellipsoidal variables in Hall (1990b), Lines *et al.* (1987, 1988), Morris (1985) and Wilson & Fox (1981).

4.3 BY Draconis variables

D. S. Hall

The BY Draconis-type variables are a subgroup of the rotating variable stars. Physically they are dKe and dMe stars – that is, late dwarfs having hydrogen line emissions in their spectra. In this case variability is produced by axial rotation of a star with non-uniform surface brightness. A region of cool spots localized on one hemisphere of the star causes the non-uniformity.

As explained in Section 6.3 on variables of the RS CVn type, BY Dra variables are one of many groups showing the phenomenon of chromospheric activity. Unlike the RS CVn binaries, however, they can be either binary or single. This fact was used to prove that occurrence in a binary is not directly responsible for chromospheric activity. Bopp & Fekel (1977) showed that an equatorial rotational velocity of 5 km s^{-1} or faster was required. The companion star in the binaries is important only indirectly, to speed up the rotation by the mechanism of spin-orbit coupling.

Several of the BY Dra variables additionally show UV Ceti-type flares, which prompts the *GCVS* to add the UV classification.

A few stars classified in the *GCVS* as BY probably should be classified FK instead, i.e., FK Comae-type variables. Though undoubtedly single, spotted, and varying as a result of rotational modulation, they are not the proper spectral type and/or luminosity class for BY variables. Examples would be OP And (gK1), V 390 Aur (K0 III), EK Eri (G8 IV-III), and V 491 Per (G8 IV).

For a historical perspective, see Hall (1994). The variability of BY Dra, the group's prototype, was discovered by P.F. Chugainov in 1966. At that time he also correctly suggested starspots to explain the variability. The variability of YY Gem was discovered long before, in 1926, but as an eclipsing binary. It is, nevertheless, clearly a BY Dra variable: its spectral type is dM1e + dM2e and its variability between the eclipses was correctly identified as a starspot wave

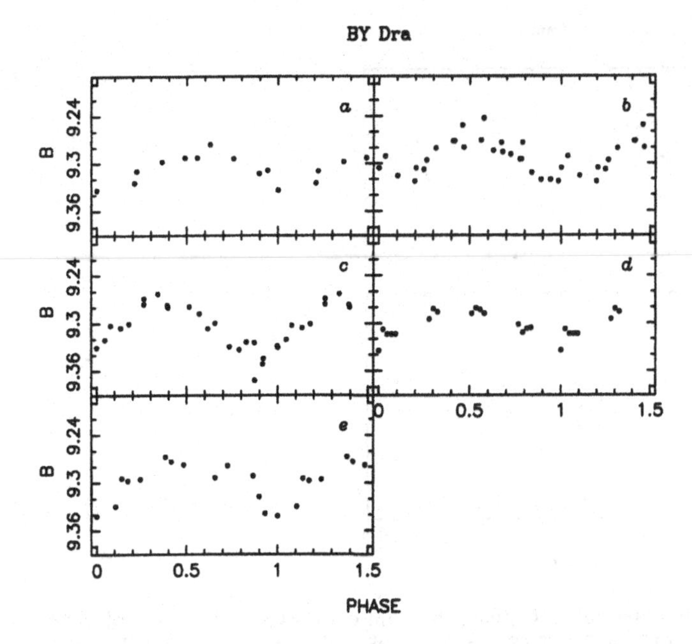

Figure 4.9 *B* phase diagrams of BY Dra (BY) (Oskanyan *et al.* 1977).

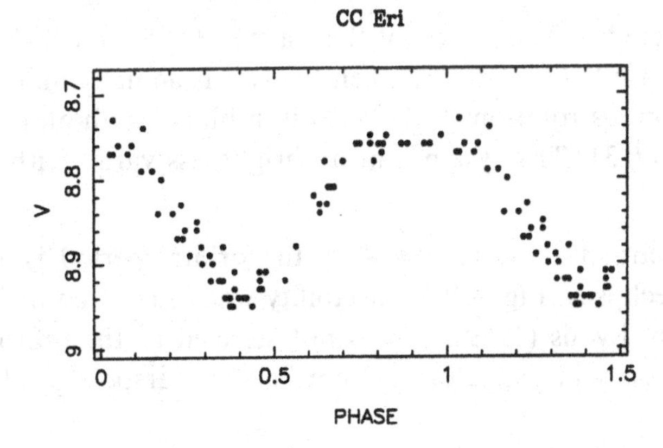

Figure 4.10 *V* phase diagram of CC Eri (BY) according to ephemeris 2430000.0 + 1.56145*E* (Evans 1959).

by Gerald E. Kron in 1952. The *GCVS*, however, emphasizing its eclipses and its flares, classifies it EA + UV.

The following three binaries are chosen to represent the BY Dra-type variables, although the group can include both binary and single stars:

(i) BY Dra: the spectral type is K4 V + K 7.5 V, the orbital period is $5^{\mathrm{d}}\!.975$, and there are no eclipses (Fig. 4.9). The light curves illustrated in the figures show starspot waves much smaller in amplitude than when

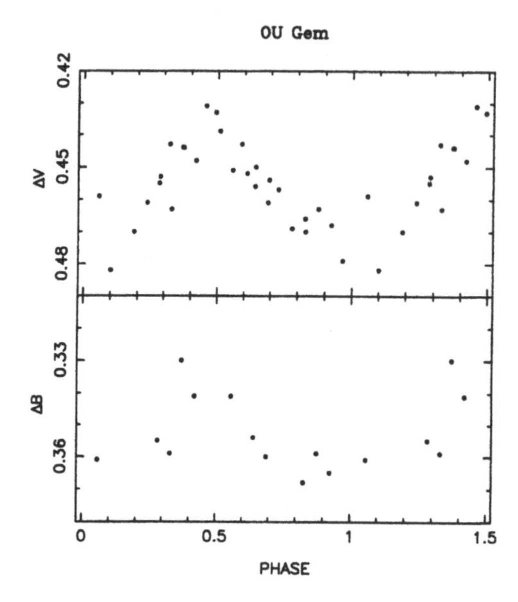

Figure 4.11 Differential *B*, *V* phase diagram of OU Gem (BY) according to ephemeris 2443846.2 + 7.36*E* (Bopp *et al.* 1981). Comparison star: HR 2277.

Chugainov discovered it. The period of the wave is $3^{d}_{\cdot}827$, much shorter than the orbital period. The ratio between the two is actually indicative of pseudo-synchronous rotation (Hall 1986) in a binary with an orbital eccentricity of e = 0.31. The system's mean brightness varies with a 50 or 60-year cycle.

(ii) CC Eri: The spectral type is K7 Ve + ?, the orbital period is $1^{d}_{\cdot}56$, and there are no eclipses (Fig. 4.10). Variability was discovered in 1959, before BY Dra, by Evans (1959). It was not selected as the prototype probably because early investigators did not believe starspots produced the variability.

(iii) OU Gem: The spectral type is K3 V + K5 V, the orbital period is $6^{d}_{\cdot}99$ days, and there are no eclipses (Fig. 4.11). The period of the wave is $7^{d}_{\cdot}36$, longer than the orbital period. This is indicative of non-synchronous rotation, which is probably a consequence of the system's relative youth.

The interested reader can learn more about chromospheric activity and the BY Dra-type variables in Baliunas & Vaughan (1985), Bopp & Fekel (1977), Byrne & Rodono (1983), Hall (1986, 1991), Linsky (1980) and Strassmeier *et al.* (1993).

4.4 FK Comae variables

D. S. Hall

The FK Comae-type variables are a subgroup of the rotating variable stars. They are rapidly rotating giants varying as a result of non-uniform surface brightness. A region of cool spots localized on one hemisphere of the star causes the non-uniformity. As originally defined by Bopp & Stencel (1981) the class included late-type giants with very high values of $v \sin i$ (short rotation period), extreme evidence of chromospheric activity (see Section 6.3 on variables of the RS CVn type), but no sign of large velocity variations (single). The *GCVS*, however, allows binaries to be included in the class.

FK Comae itself rotates so rapidly that the most reasonable evolutionary scenario involves the coalescence of a W UMa-type binary and a surrounding optically thick spun-up envelope. Other stars assigned to this class do not rotate so rapidly and may be simply evolved single A-type stars which have not lost much of their original rapid main-sequence rotation. If binaries are allowed in the class, then their rapid rotation will be a result simply of synchronization with a rather short orbital period.

Only four stars classified FKCOM appear in the Fourth Edition of the *GCVS*, and only a few more have appeared in subsequent Name Lists; several stars not classified as FKCOM probably should have been. The single K1 giant OP And = HR 454 (P = 37 days) has been assigned to the BY Dra class, which includes only dwarfs. The single G8 giant KU Peg = HD 218153 (P = 22 days) has been assigned to the RS CVn class, which includes only binaries.

The prototype FK Com is the most rapidly rotating of all, with a mean period of $2^{d}.4$. Among all variables in the class, measured amplitudes have ranged from a few hundredths to a few tenths of a magnitude.

The following two variables are chosen as representative of the class:

(i) UZ Lib: the spectral type is K2 III and the unseen companion is probably a dwarf M star. The orbital period is $4^{d}.768$ and the photometric period is almost the same: $4^{d}.75 \pm 0^{d}.01$ (Fig. 4.12). Starspots produce a very large light variation, which has been as large as $0^{m}.35$ magnitude in *V*.

(ii) OU And: the spectral type is G1 III and the giant is almost surely single. Variability due to starspots defines a rotation period of approximately 23 days but only a small amplitude, not more than $0^{m}.04$ in *V* (Fig. 4.13). OU And lies in the Hertzsprung gap, probably having evolved from a rapidly rotating single A-type main-sequence star but not yet having

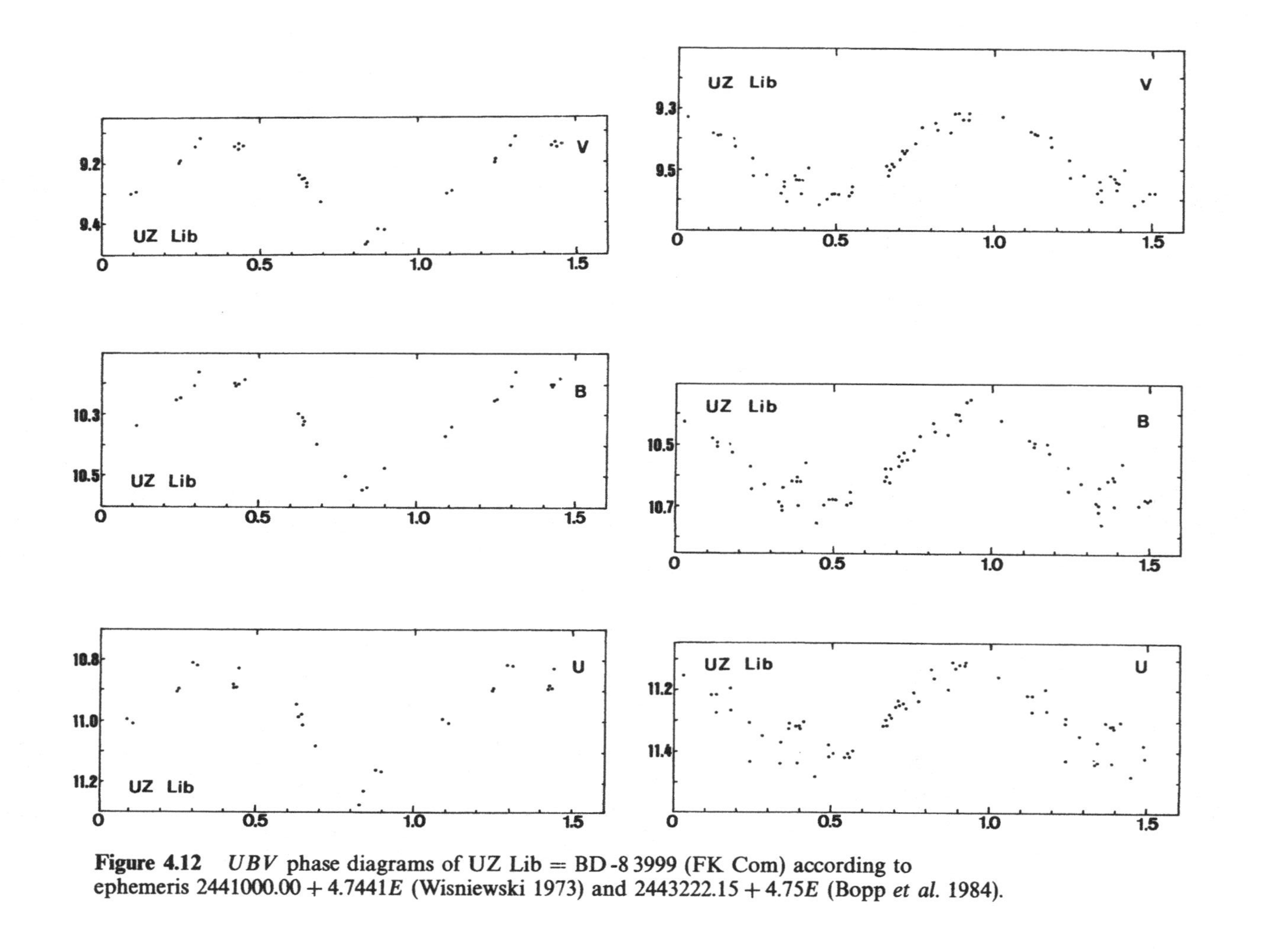

Figure 4.12 *UBV* phase diagrams of UZ Lib = BD -8 3999 (FK Com) according to ephemeris 2441000.00 + 4.7441E (Wisniewski 1973) and 2443222.15 + 4.75E (Bopp *et al.* 1984).

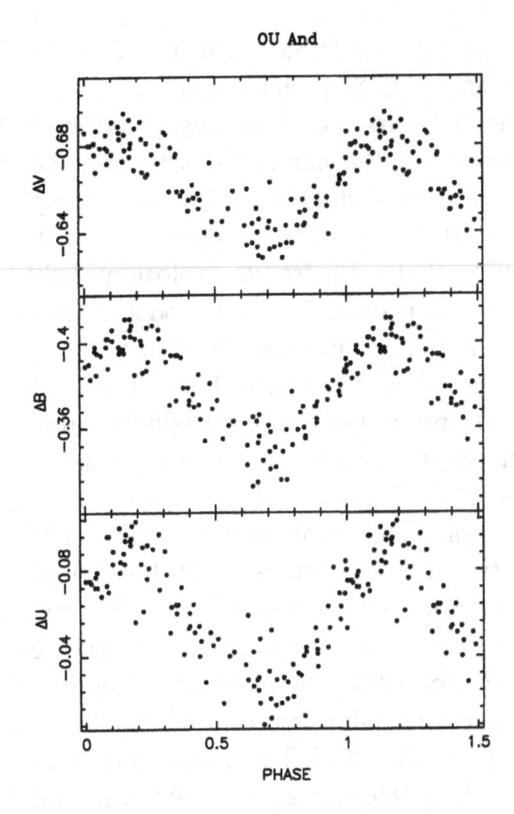

Figure 4.13 Differential *UBV* phase diagrams of OU And = HR 9024 (FK Com) according to ephemeris 2446311.0 + 22.50*E* (Boyd *et al.* 1990). Comparison star HD 223848.

passed the rotation break which Gray (1989) says occurs between G0 III and G3 III.

The interested reader can learn more about the FK Comae-type variables, and about rapidly rotating giant stars in general, in Bopp & Stencel (1981), Fekel *et al.* (1986), Gray (1989) and Strassmeier & Hall (1988).

4.5 Pulsars

J. Krautter

Pulsars are rapidly rotating neutron stars which emit very regular pulses with periods between 1.558 ms and 4.308 s. Most pulsars are found on the basis of their radio emission, only very few pulsars also emit pulses in the optical spectral range. For that reason they are usually called radiopulsars. (One should not confuse radiopulsars with X-ray pulsars which emit pulses in the X-ray regime; they belong to the group of X-ray binaries.)

The possible existence of neutron stars was first postulated in 1932 by Landau – shortly after the discovery of the neutron – who speculated that stars could consist of neutron matter. Such stars should be very compact and small. Independently from Landau, Baade and Zwicky postulated, in 1934, the existence of neutron stars, assuming that a supernova explosion could be the transition from ordinary stars to neutron stars. The first model of the possible structure of a neutron star was published by Oppenheimer and Volkoff in 1939. They found that neutron star masses should not exceed 1 to 2 $M_\odot$. However, because of their rather extraordinary properties, and especially because of their small size, it was not believed that neutron stars could be ever observed.

The discovery of radiopulsars in 1967 came rather unexpectedly, even if Pacini (1967) had shortly before predicted that neutron stars should emit strong non-thermal electromagnetic radiation. The team of discoverers, led by Anthony Hewish from Cambridge University (England), was investigating – at high time resolution – scintillating radio sources using an array of radio antennas completed during the fall of 1967. In November, Jocelyn Bell, a graduate student, detected from one particular direction of the sky extremely regular radio pulses with a period $P = 1\overset{s}{.}3373011$. The nature of this first pulsar CP 1919 – now called PSR 1919+21 – (Hewish *et al.* 1968) remained highly speculative for about a year, even after several other pulsars were subsequently detected in 1968. (Since the pulses arrived with such a striking regularity, it was soon speculated whether they could be signals from extraterrestrial intelligent living beings.) The neutron star nature was settled at the end of 1968 by the detection of a pulsar in the middle of the Crab nebula (PSR 0531+21), the remnant of a supernova which exploded in 1054 AD. Now more than 500 pulsars are known with periods in the range $0\overset{s}{.}001558 \leq P \leq 4\overset{s}{.}308$.

Neutron stars are final products in the evolution of a star; they are formed in the event of a supernova II explosion by the collapse of the iron core of the SN progenitor. A second mechanism for the formation in an SN I explosion due to the collapse of an accreting white dwarf is still highly debated. The rapid rotation of the neutron stars is a direct consequence of the conservation of the angular momentum during the collapse phase. The angular velocity increases inversely proportional to the square of the stellar radius by about a factor of 10^{10}. Since the magnetic field is 'frozen' in the stellar matter during the collapse, a similar amplification of the magnetic field takes place resulting in field strengths of about 10^9 tesla ($\equiv 10^{13}$ gauss).

Many details of the emission mechanism of the pulsed radiation are still unclear, however, the basic processes now seem to be understood. Prerequisites are both the rapid rotation and the strong magnetic fields. If the rotation axis and the axis of the magnetic field are tilted with respect to each other, huge electric fields will be created in the immediate neighbourhood of the pulsar. These fields are accelerating the free electrons and protons present at the surface of the neutron star to relativistic velocities. This plasma moves away from the neutron star along the magnetic field lines, which are not closed at the polar region. The relativistic electrons emit (non-thermal) synchroton radiation in a narrow cone, comparable to the beam of a lighthouse. This cone is rotating with the angular velocity of the pulsar; a pulse is received if the beam intercepts the line of sight of an observer. This implies that there must be many pulsars whose beams never intercept the line of sight towards the earth, and hence, which are hidden from us.

The energy of the synchrotron radiation comes from the rotational energy of the spinning neutron star. Although the details are still poorly understood, there is a constant conversion of rotational energy into the radiant energy of the synchrotron radiation. Observations have shown that without exception the periods of all pulsars are slowing down at a rate which is proportional to the dissipated energy. The period changes are between 10^{-18} and 10^{-12} $s\,s^{-1}$. The observed rate of the period increase of the Crab pulsar of about 4.2 10^{-13} $s\,s^{-1}$ corresponds to the total energy radiated away by the synchrotron radiation. According to this scenario the pulsars with the fastest rotation should be the youngest ones, which has indeed turned out to be true in the case of the Crab pulsar (Fig. 4.14). With increasing period the energy output decreases, and after a certain time of several million years the emitted radiation becomes too faint to be measured any longer. An exception is the so called 'millisecond' pulsars which have extremely short periods, but which are believed to be very old objects. It is now assumed that the millisecond pulsars were formed in close binary systems. The speed up of the rotational velocity is due to transfer of matter from a 'normal' star to the pulsar. Occasional decreases of the pulsar periods, the so-called 'glitches', which are due to changes in the internal structure of the pulsar, do not affect the general picture described above.

For most radio pulsars the duty cycle is of the order of several percent. Individual radio pulses from a particular pulsar show a variety of intensity variations, and, in part, also some microstructure within a pulse. However, the superposition of many pulses gives for each individual pulsar a rather stable and characteristic pulse profile. Some of the pulses extend over the major part of the pulsar rotation period or even over the total period. Many pulsars exhibit profiles which show two distinct components of different strength representing two beams radiated from the two opposite magnetic poles.

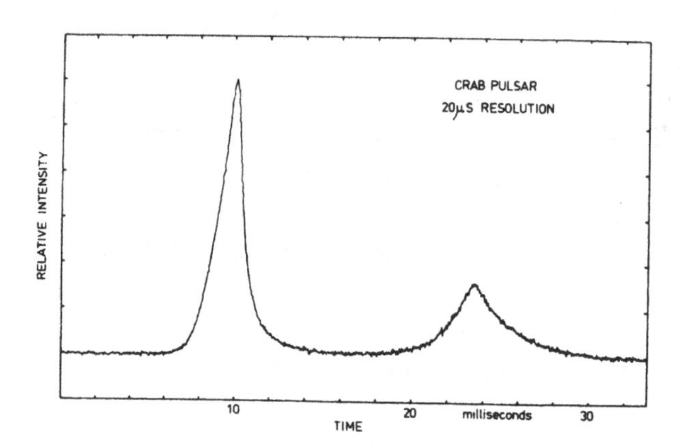

Figure 4.14 Optical light curve of the Crab pulsar (= CM Tau) obtained by Smith *et al.* (1978) in white light with a very high time resolution of 20 μs.

Radiation in the optical spectral range could be observed only in the case of three pulsars, the Crab pulsar (PSR 0531+21, P = 33.3 ms), the Vela pulsar (PSR 0833-45, P = 89.3 ms), and a pulsar in the LMC (PSR 0540-69, P = 50.3 ms). This is due to the fact that the luminosity of the optical radiation (and of the radiation at higher energies) shows a very strong dependence on the orbital period, which is – according to Pacini (1967) – approximately proportional to P^{-10}. For instance, the time-averaged luminosity of the optical pulses from the Vela pulsar is approximately five orders of magnitude smaller than the corresponding luminosity from the Crab pulsar. It is therefore not surprising, that only the optical light curve of the Crab pulsar has hitherto been measured with reasonable signal-to-noise. The first discovery of optical pulses from the Crab pulsar was made less than one year after its radio detection by Cocke *et al.* (1969). Figure 4.14 shows the optical light curve of the Crab pulsar (≡ CM Tau) obtained by Smith *et al.* (1978) in white light with a very high time resolution of 20 μs. The optical light curve coincides very well with both the radio and the X-ray light curve showing a main and a secondary pulse.

For more general literature, we refer to Helfand *et al.* (1985), Taylor & Stinebring (1986), Srinivasan (1989), Lyne & Graham Smith (1990) and Ventura & Pines (1991).

5

Cataclysmic (explosive and nova-like) variables

5.1 Supernovae

H.W. Duerbeck

A supernova explosion is a rare type of stellar explosion which dramatically changes the structure of a star in an irreversible way. Large amounts of matter (one to several solar masses) are expelled at high velocities (several tens of thousands km s^{-1}). The light curve in the declining part is powered by thermalized quanta, released by the radioactive decay of elements produced in the stellar collapse, mainly ^{56}Co and ^{56}Ni. The ejected shell interacts with the interstellar medium and forms a SN remnant, which can be observed long after the explosion in the radio, optical and X-ray regions.

Supernovae can be divided into two classes (and several subclasses), viz. SN I and SN II.

SN I have fairly similar light curves (see, for example, Fig. 5.1) and display a small spread in absolute magnitudes. Spectra around maximum show absorption lines of Ca II, Si II and He I, but lack lines of hydrogen. They occur in intermediate and old stellar populations. Their progenitor stars are not clearly identified, but massive white dwarfs (WDs) that accrete matter from a close companion and are pushed over the Chandrasekhar limit are good candidates. Another possibility is the hypothesis of the fusion of a binary consisting of two WDs. The collapse of the white dwarf leads in both cases to an explosive burning of its carbon, and the released energy is sufficient to trigger a disintegration of the complete object.

Recent studies have found evidence for a subdivision of type I into Ia, Ib and Ic. The light curves in these three subclasses are practically identical. Spectroscopically, the Si II-absorption at 615 nm is missing in subtype b, and the absolute magnitude of subtype b objects at maximum is approximately $1^{\mathrm{m}}.5$ below that of type a. Type Ib objects belong to population I (contrary to type

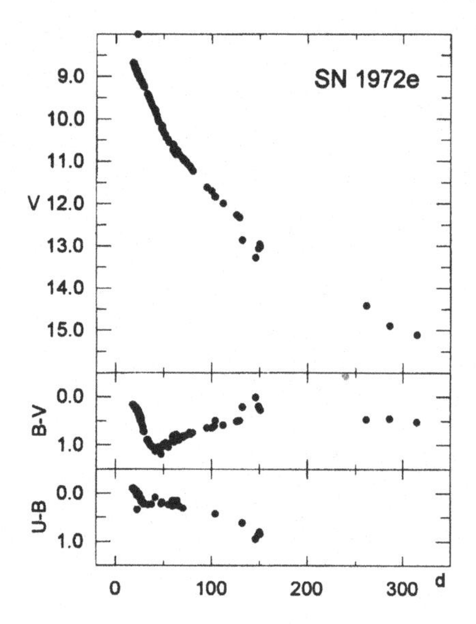

Figure 5.1 *UBV* light and colour curves for SN 1972e (SN I) from Ardeberg & de Groot (1973). The X-axis is Julian days since the moment of maximum brightness (JD 2441438).

Ia objects, which definitively belong to population II). It is possible that the disintegration of a white dwarf triggers both SN Ia and Ib, with the difference that in the latter case the event takes place in a massive, hydrogen-poor star, e.g. a Wolf–Rayet star, and part of the energy generated is not available in the form of radiation, but is used to eject the outer envelope. If the spectra around maximum show neither Si II 615 nor He I 587.6, the supernova is classified as Ic. Objects of subtype Ic are probably similar to those classified as Ib.

SN II have light curves that show a greater diversity, the spread in absolute magnitude is larger than for type I (Figs. 5.2–5.5) . The cause of the outburst is the exhaustion of stellar energy sources when an iron core is formed. The star collapses and forms a neutron star, and the material following its collapse bounces at the core and is ejected, together with the outer layers; neutrino pressure may also play a role in the early stages of the explosion. Spectra around maximum light show blueshifted absorption lines of H and He I, the later emission line spectrum is not unlike that of novae, but has higher expansion velocities. SN II occur in young stellar populations; they constitute the final stage of massive single stars following the red (rarely blue) supergiant stage, leading to neutron-star or black-hole remnants.

Supernovae have been known since antiquity; Chinese and Korean lists of guest stars contain a certain percentage of Galactic supernovae, like SN Lupus (SN 1006, type Ia) and CM Tau (SN 1054, type II). The last observed

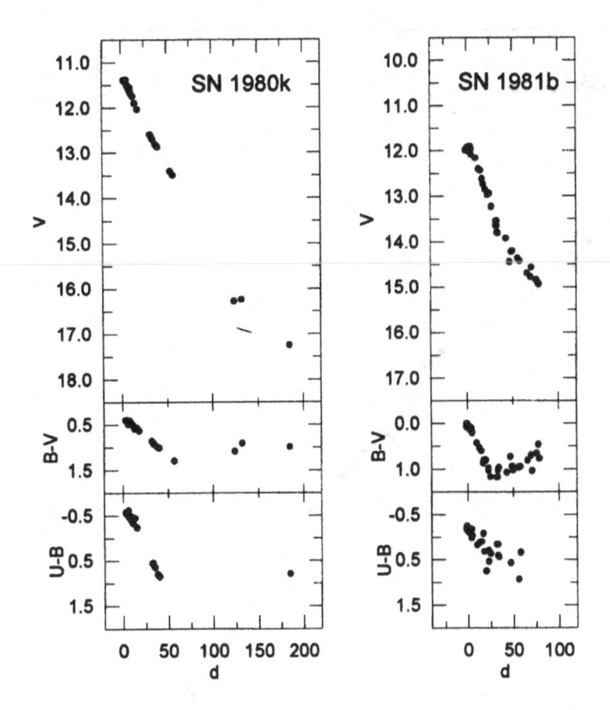

Figure 5.2 *UBV* light and colour curves for SN 1980k (SN II L) from Buta (1982) and for SN 1981b (SN I) from Buta & Turner (1983). The X-axis is in Julian days since the moment of maximum brightness (JD 2 444 544 for SN 1980k and JD 2 444 672 for SN 1981b).

Galactic SN were Tycho's SN 1572 (B Cas, type Ib?) and Kepler's SN 1604 (V 843 Oph, type Ib/II?), while the most recent one was perhaps overlooked (Cas A, about 1658, type Ib). The *GCVS* lists three of the above objects plus a few extragalactic ones which were included only by chance because their extragalactic nature was not yet known. The nearest extragalactic supernovae are S And (1885, in M31, type still under discussion) and SN 1987a (in the LMC, type II). Hundreds have been discovered in more distant galaxies, especially since the pioneering search by F. Zwicky and collaborators, which started in 1936 (Zwicky 1974).

Absolute magnitudes of supernovae at maximum light are: $M_V = -19.9$ for type Ia, $M_V = -18.0$ for type Ib, and $M_V = -17.8$ for type II, with an assumed Hubble-parameter of 50 km $s^{-1}Mpc^{-1}$ (derived from Barbon *et al.* 1989).

Frequencies of occurrence of supernovae are dependent on the SN type and the type of galaxy, as well as on the luminosity of the galaxy. The discovery is influenced by the absolute brightness of the SN, the dust absorption, the inclination, etc. of the host galaxy, and the derivation of true frequencies is

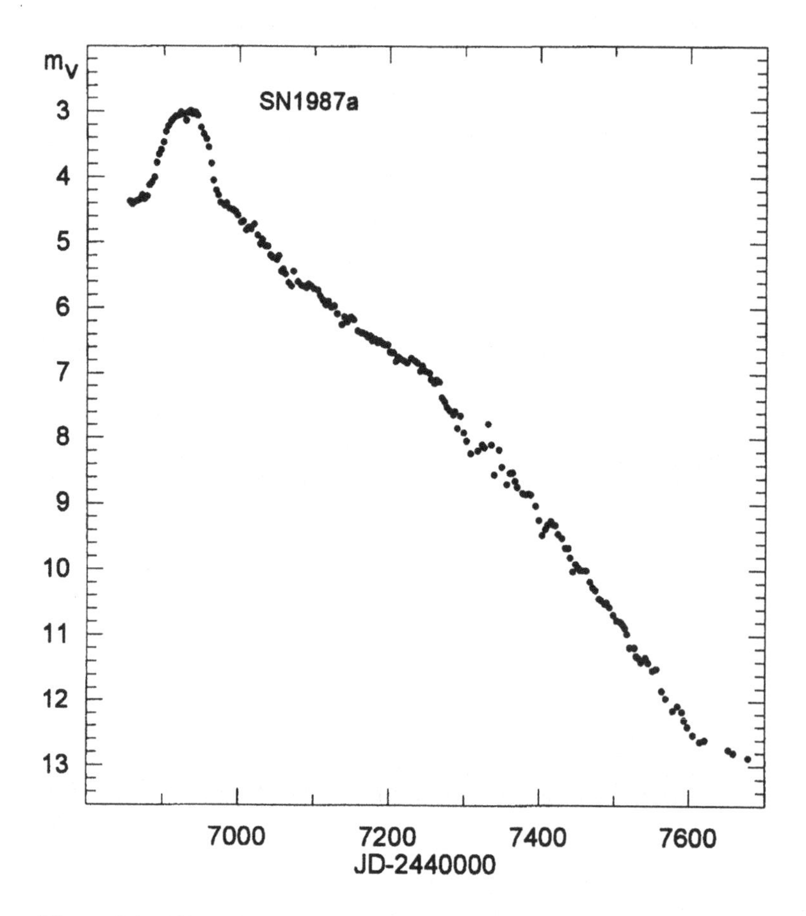

Figure 5.3 Visual light curve for SN 1987a (AAVSO).

still controversial. In Table 5.1, the SN frequencies are given in SN-units (1 SN-unit equals 1 supernova per century per 10^{10} $L_{\odot}$(B), h is the value of the Hubble-parameter in units of 100 km $s^{-1}Mpc^{-1}$). These values have been averaged over recent determinations made by different authors.

Table 5.1 Supernova frequencies for different types of galaxies

Type	E-S0	S0/a,Sa	Sab,Sb	Sbc-Sd	Sdm-Im
Ia	0.5 h^2	0.5 h^2	0.35 h^2	1.3 h^2	0.4 h^2
Ib	—	0.15 h^2	0.25 h^2	1.05 h^2	0.7 h^2
II	—	0.25 h^2	1.0 h^2	6.0 h^2	2.5 h^2
all	0.5 h^2	0.9 h^2	1.6 h^2	8.3 h^3	3.6 h^2

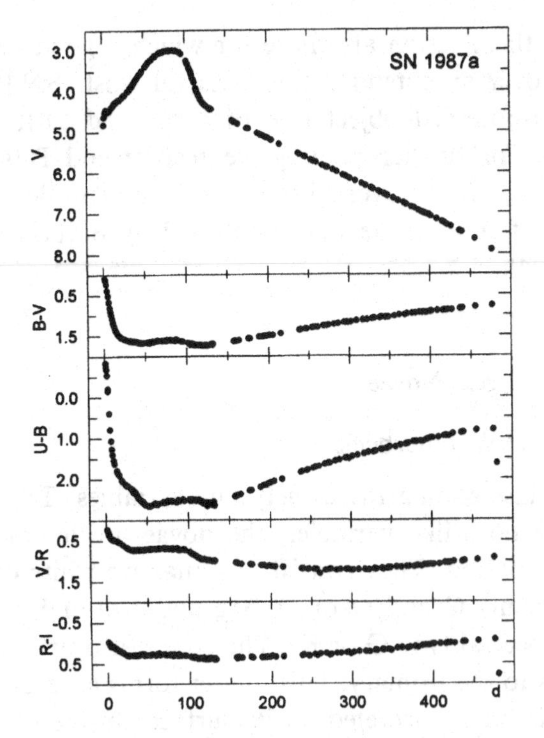

Figure 5.4 *UBVRI* light and colour curves for SN 1987a (SN II pec), from Hamuy *et al.* (1988). The X-axis is in Julian days since the moment of core collapse (JD 2 446 849).

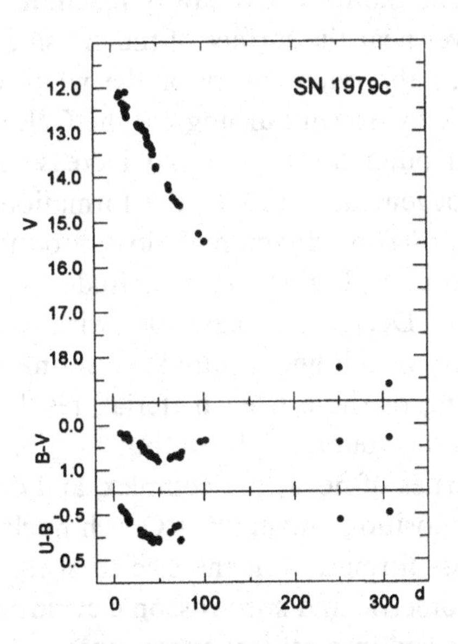

Figure 5.5 *UBV* light and colour curves for SN 1979c (SN II) from de Vaucouleurs *et al.* (1981). The X-axis is in Julian days since the moment of maximum brightness (JD 2 443 979).

The selected light curves in this Section are those for which a good record of photoelectric observations over an extended time interval exist. SN 1987a (Type II) is certainly the best-observed object (see Figs. 5.3 and 5.4); it is, however, somewhat atypical. For further reading we recommend Petschek (1990), Woosley (1991), Woosley & Weaver (1986) and van den Bergh & Tammann (1991). Catalogues of supernovae were published by Barbon *et al.* (1989) and van den Bergh (1994).

5.2 Novae

H.W. Duerbeck

Novae (N) are an important class among the cataclysmic variables. Together with the dwarf novae and the nova-like variables, the novae are interacting binary stars, generally of short period. They consist of a massive white dwarf (the primary, of chemical elements C and O or O, Mg and Ne) and a cool dwarf star (the secondary, of spectral type G or K). The cool star overflows its critical volume, and loses mass to the primary. This matter forms an accretion disc around the primary and is finally accreted on its surface. Instabilities in the accretion disc lead to short and long period photometric variability at the stage of minimum light.

The cause of the nova outburst is a thermonuclear runaway reaction, which occurs in the accreted hydrogen-rich layer near the surface of the massive white dwarf into which C and O nuclei from the outer layers of the white dwarf are mixed. At a certain critical pressure, hydrogen burning via the CNO cycle starts in this degenerate hydrogen-rich outer layer. A rapid increase of the temperature leads to a lifting of the degeneracy, and to the formation of a shock wave. This, in combination with radiation-driven mass-loss, produces an expanding atmosphere of large size and of high absolute magnitude (typically $M_V = -6$ to -9) at maximum light. Decreasing mass loss with ongoing energy release causes a decline of the visual light output, a shrinking of the photosphere, and a radiative heating of the ejected material, resulting in interesting spectroscopic phenomena in the course of the outburst.

The detailed light and spectral properties of novae are complex, and depend on white dwarf mass and chemical composition, mixing of CO-rich nuclei into the accreted material, as well as on dust formation in the ejected shell. Each nova has its unique, characteristic photometric and spectroscopic evolution. In spite of this, novae can be broadly classified into several subgroups:

(i) NA – fast novae which, after maximum light, decline three magnitudes in visual light in 100 days or less. They have usually fairly smooth light curves and generally have higher absolute magnitudes.

(ii) NB – slow novae which decline three magnitudes from maximum in visual light in more than 100 days. They have usually fairly structured light curves and, as a rule, fainter absolute magnitudes.

Members of the groups NA and NB are also called *classical novae.* Their absolute magnitude M_V is correlated with the fastness of the light-curve decay: faster novae are more luminous at maximum light. The fastness is measured by the t_2- or t_3-time: a nova takes a certain time (measured in days) to decline by 2 or 3 magnitudes from maximum light. $M_V - t_3$-time calibrations have been carried out by several authors. For a summary of earlier relations, see Duerbeck (1981), and for a more recent calibration, Cohen (1985).

Novae have been known since antiquity; Chinese and Korean lists of guest stars contain a certain percentage of novae, which unfortunately cannot be identified at minimum today due to low positional accuracy. The oldest identified novae are CK Vul (1670), WY Sge (1783) and V 841 Oph (1848); Q Cyg (1876) is the first classical nova for which spectroscopic observations exist.

By the end of 1986, about 215 novae had been discovered in the Galaxy, systematic searches by Arp and Rosino and collaborators led to the discovery of 142 novae in M31, the Andromeda galaxy (more have been found by earlier searches and in more recent, specialized studies), 21 in the Large Magellanic Cloud and 6 in the Small Magellanic Cloud. See Cappacioli *et al.* (1989, 1990) for a survey of novae in M31 and the Large Magellanic Cloud.

(iii) NC – very slow novae, which remain near maximum light for years or even decades. The bulk of these objects are symbiotic stars, accreting objects with late-type giant companions. They are often called *symbiotic novae.* (The reader is referred to Section 5.5 on Z And stars).

(iv) NR – recurrent novae. These novae show repeated outbursts at time intervals of decades, in contrast to classical novae which are known to show only one eruption in historic times (this also means that there exists a population of close binaries which are classical novae but which have not yet undergone an explosion in the last centuries, and will remain in a low state for the next centuries or millenia. Some of these objects are probably included in the class of nova-like variables, see Section 5.3).

Recurrent novae are usually fast novae (t_3-times of a few days), often have giant companions, and the accreting white dwarfs are probably near

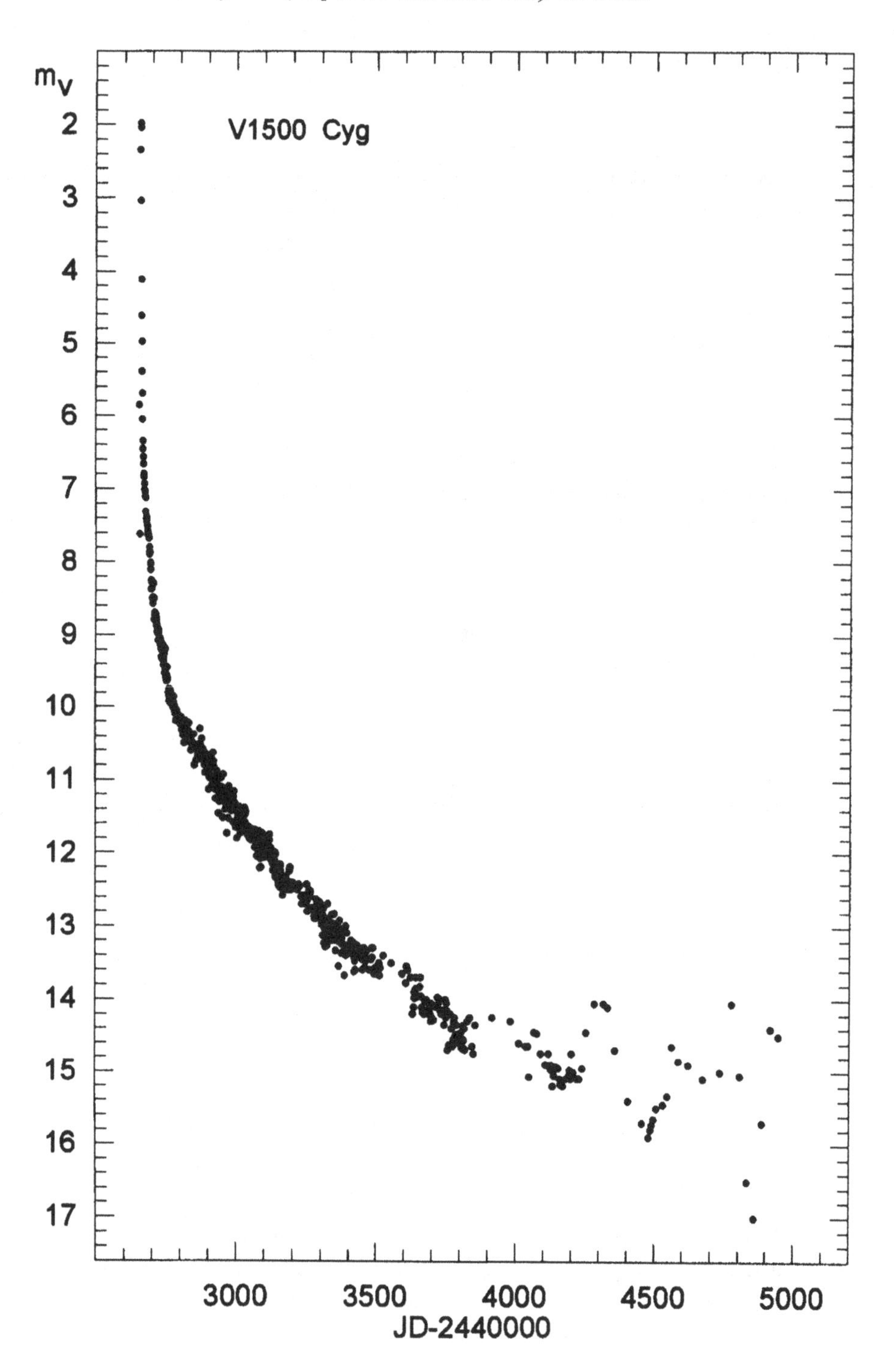

Figure 5.6 Visual light curve for V 1500 Cyg (NA = fast nova), data from AAVSO.

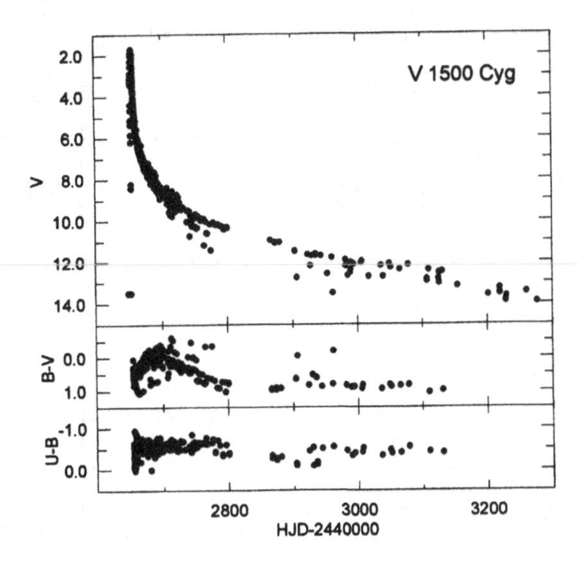

Figure 5.7 *UBV* light and colour curve for V 1500 Cyg (NA = fast nova), based on an unpublished compilation of photoelectric observations by H.W. Duerbeck.

the Chandrasekhar limit, a position that allows explosions in degenerate material under high pressure. The 'critical' amount of accreted material, which undergoes the thermonuclear runaway reaction, is about a factor 10 less than in the case of classical novae. NR appear somewhat underluminous in the $M_V - t_3$-time diagram. So far, only eight objects have been discovered in the Galaxy, two in M31, and one in the Large Magellanic Cloud.

The first recurrent nova discovered as a nova was T CrB (1866, 1944); recurrency was first found (by means of the study of the Harvard Plate collection) for T Pyx in 1902.

We have selected for display a number of light curves of recent novae for which extensive *UBV* observations were made, and that cover novae of different speed classes without and with dust formation. The fastest nova on record is V 1500 Cyg (1975), see Figs. 5.6 and 5.7. FH Ser (1970) (Figs. 5.8 and 5.9) is a famous nova with dust formation, in which the 'constant luminosity' phase was discovered: while the visual brightness declined, the ultraviolet brightness increased (this can also be seen in the colour indices), and when the dust formation set in, the visual and the ultraviolet brightness dropped dramatically, while an infrared excess became prominent. HR Del (1967) (Figs. 5.10 and 5.11) is a good example of a slow nova with a very long pre-maximum stage.

The colour indices of novae are largest near maximum light, when the expanding photosphere of a nova reaches its largest extent. While ejected

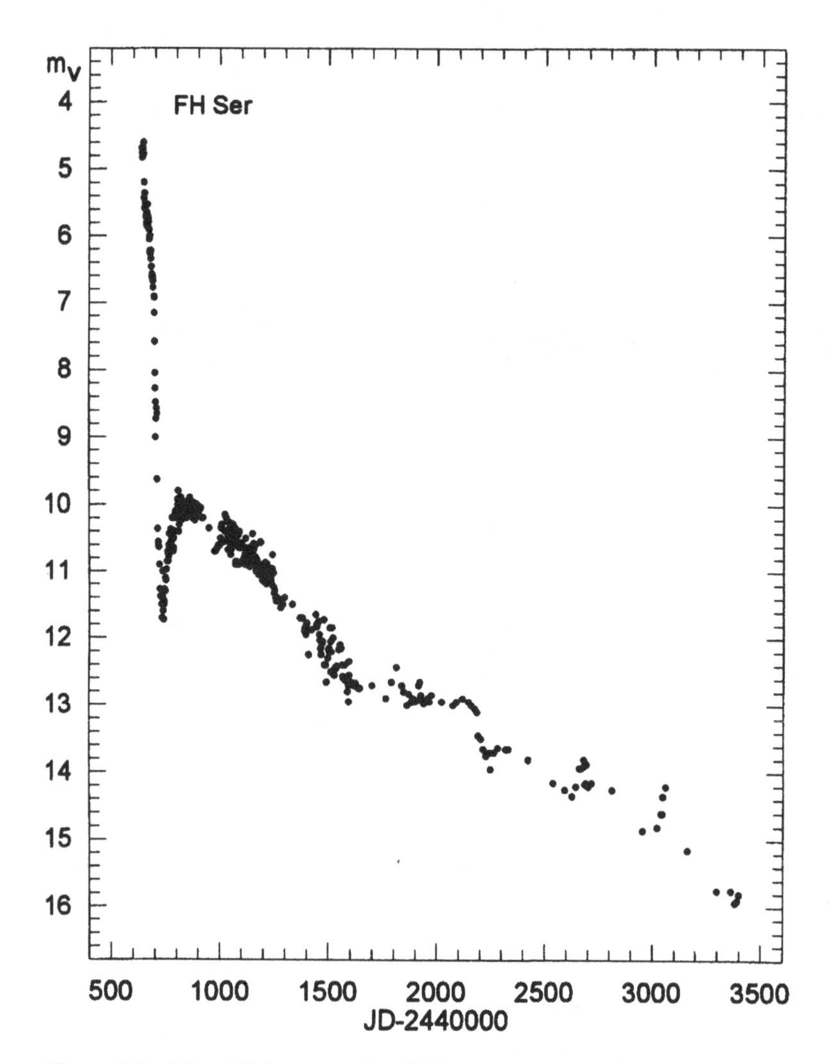

Figure 5.8 Visual light curve for FH Ser, a nova with dust formation. Data from AAVSO.

material continuously flows outward, the photosphere contracts and heats up, making the colours bluer. The spectra of novae in late phases are dominated by emission lines, and broad-band magnitudes and colours determined by different observers using filters with slightly different transmission curves deviate markedly, as can be seen in the light curves which are based on the published data of many observers.

There is practically no recurrent nova with a good record of photoelectric observations. The best-observed outburst is that of RS Oph (see Fig. 5.12).

A thorough data collection and analysis of novae observed till about 1950 is given in Payne-Gaposchkin (1957), a modern overview was published by

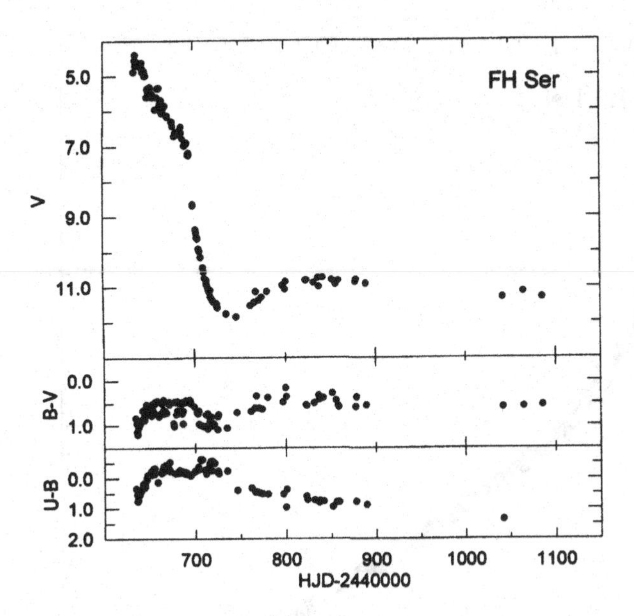

Figure 5.9 *UBV* light and colour curve for FH Ser, a nova with dust formation (based on an unpublished compilation of photoelectric observations by H.W. Duerbeck).

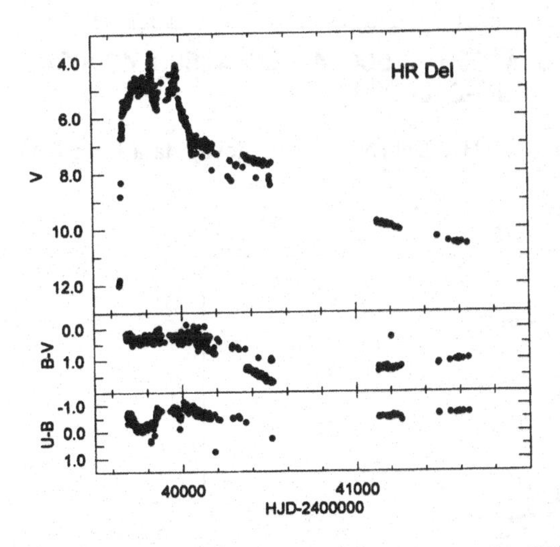

Figure 5.10 *UBV* light and colour curve for HR Del (NB = slow nova), based on an unpublished compilation of photoelectric observations by H.W. Duerbeck.

Bode & Evans (1989). A collection of physical data, positions, finding charts and bibliographic information on all novae up to 1986 was compiled by Duerbeck (1987). Many review articles cover the complete field of nova research (Gallagher & Starrfield 1978, Gehrz 1988, Shara 1989); the most

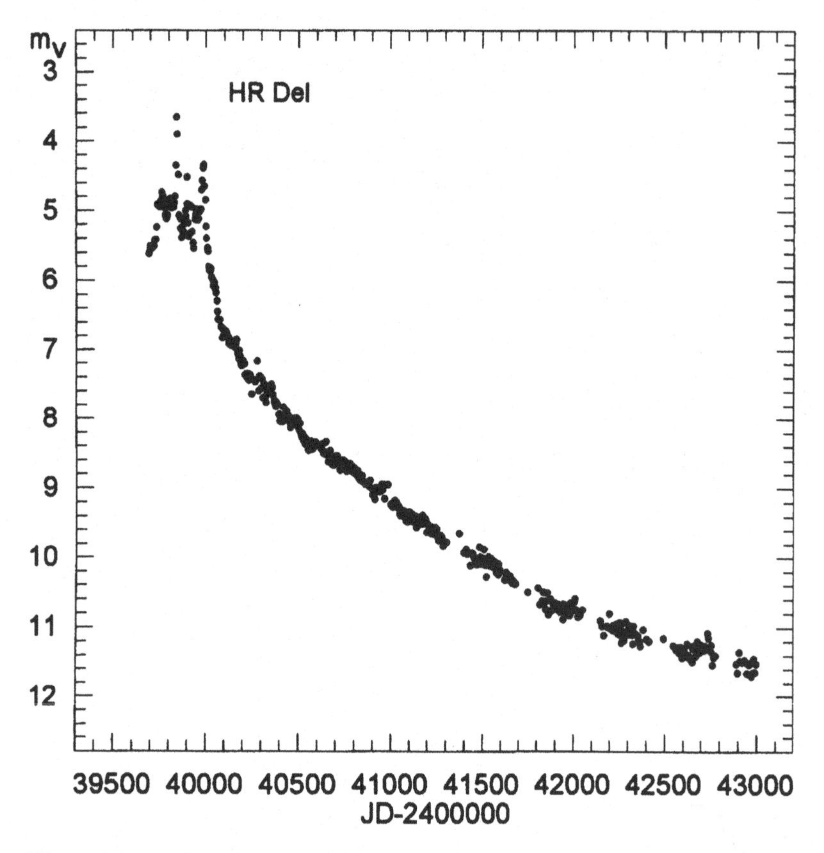

Figure 5.11 Visual light curve for HR Del (NB = slow nova), data from AAVSO

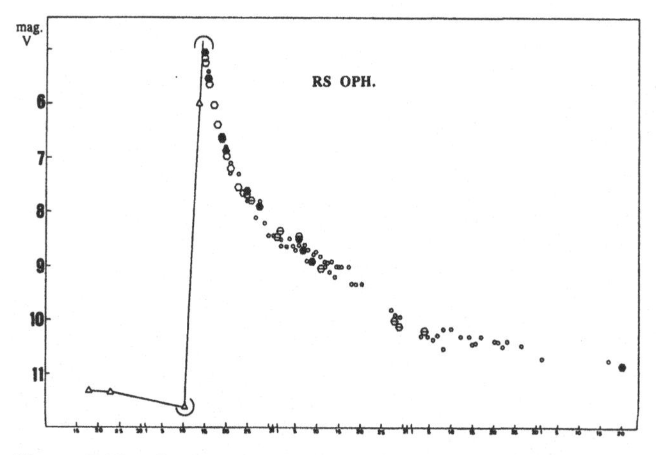

Figure 5.12 Outburst of RS Oph, graph based on *UBV* data from Connelley & Sandage (1967), Cousins (1958) and Rosino *et al.* (1960).

recent IAU Colloquium dedicated to novae was edited by Cassatella & Viotti (1990).

5.3 Nova-like stars

H.W. Duerbeck

This group is classified in the *GCVS* as 'insufficiently studied objects resembling novae by the characteristics of light changes or by spectral features. This type includes, besides variables showing nova-like bursts, also objects with no bursts ever observed; the spectra of nova-like variables resemble those typical for old novae at minimum light. Quite often a detailed investigation makes it possible to reclassify some representatives of this highly inhomogeneous group into some other type of variable star.'

We do not know exactly how novae look for the long interval between outbursts – according to the hibernation hypothesis, accretion can be dramatically reduced, and novae may then not look like nova-likes. If the accretion rate and the magnetic field strength of the white dwarf are low enough, quasi-periodic disc instabilities can occur, and the object is then classified as a dwarf nova. If the white dwarf mass is high enough ($M_{WD} > 0.6M_{\odot}$), nova explosions can occur, and the object is classified as a nova, *if* such an event has occurred in the last decades and was properly recorded. In all other cases, i.e. when signatures of accretion on the white dwarf via a disc or an accretion column are present in the spectrum, and the object cannot be clearly classified as N or DN, it is classified as NL.

Ritter (1990) uses the following subdivisions of NL:

(i) AC = AM CVn systems. These systems do not contain hydrogen, and are presumably composed of two white dwarfs. Light variations are caused by rotational, pulsational and accretion effects (flickering). Orbital periods are extremely short (4 objects).

(ii) AM = AM Her systems (polars). These systems contain a synchronously rotating, magnetized white dwarf (magnetic moment of $\approx 10^{34}$ gauss cm^3) and a cool companion which is near the main sequence. The accretion occurs towards the magnetic poles, the systems show polarized optical radiation, strong X-ray radiation, short period modulations (orbital motion effects) as well as long-term bright and low states. Orbital periods are below 3.5 hours (17 objects).

(iii) DQ = DQ Her systems (intermediate polars, IP). These systems contain a non-synchronously rotating, magnetized white dwarf (magnetic moment ≈ 0.1 times as strong as in the AM systems) and a cool companion

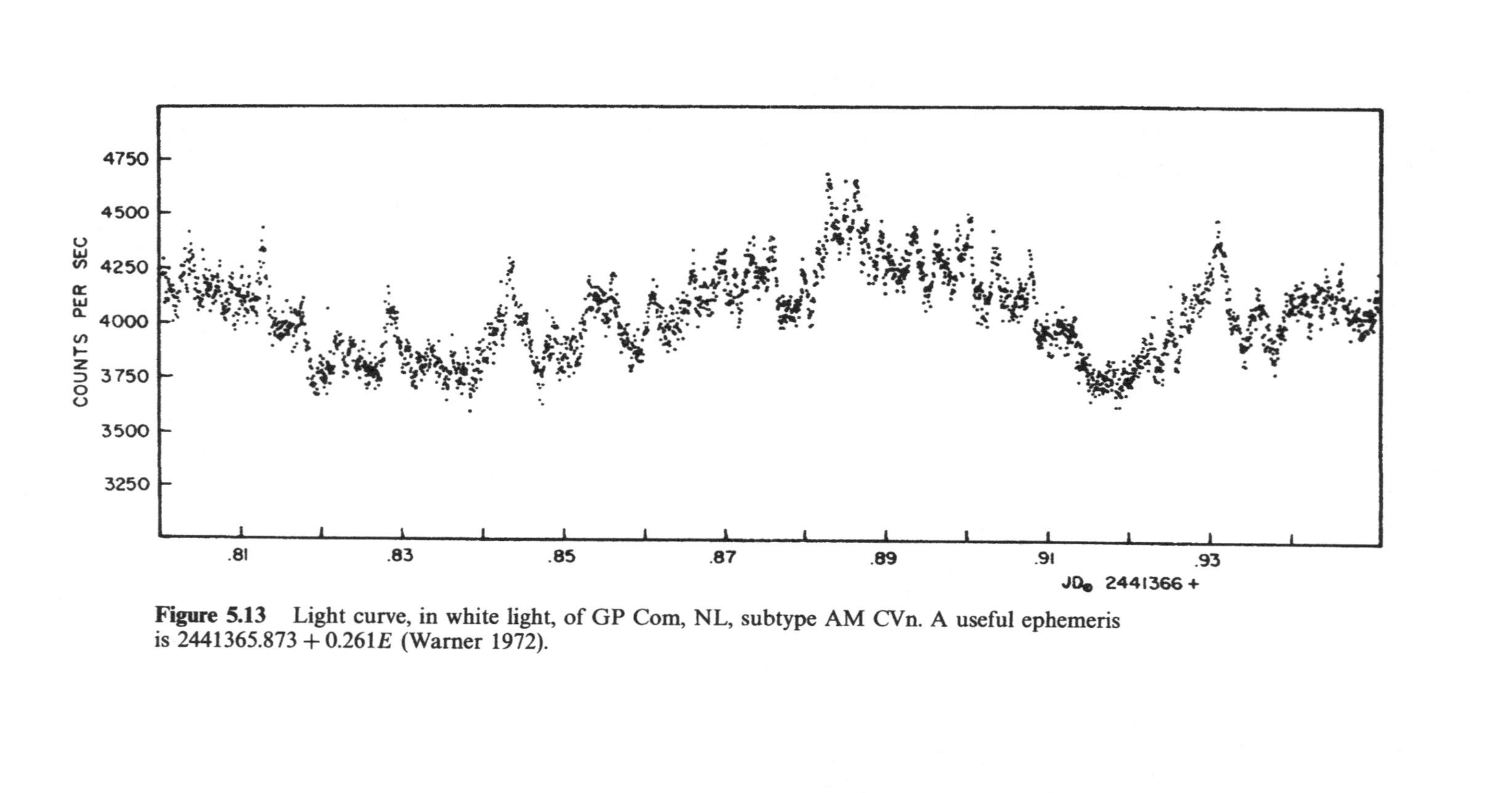

Figure 5.13 Light curve, in white light, of GP Com, NL, subtype AM CVn. A useful ephemeris is $2441365.873 + 0.261E$ (Warner 1972).

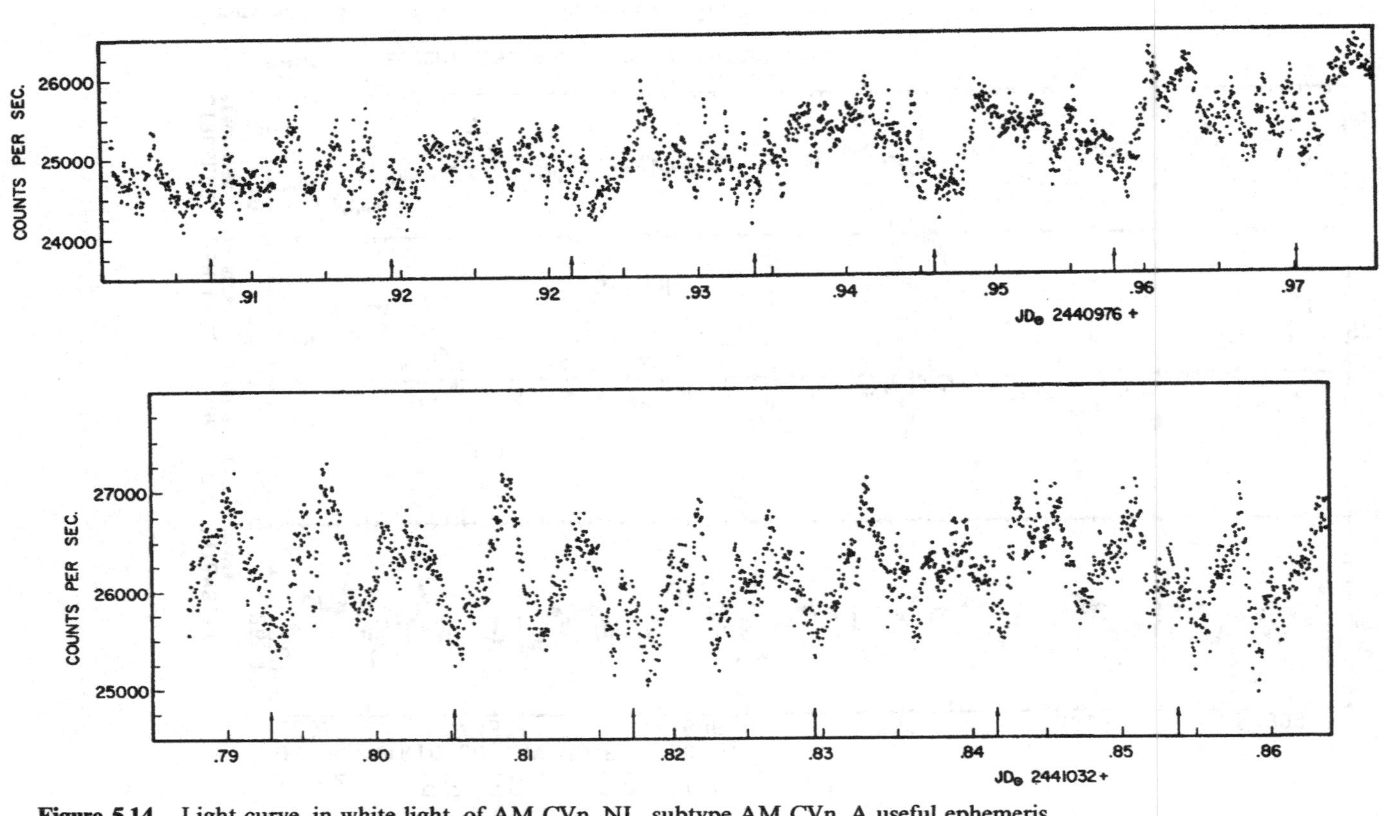

Figure 5.14 Light curve, in white light, of AM CVn, NL, subtype AM CVn. A useful ephemeris is $2441032.8052 + 0.012164936E$ (Warner & Robinson 1972).

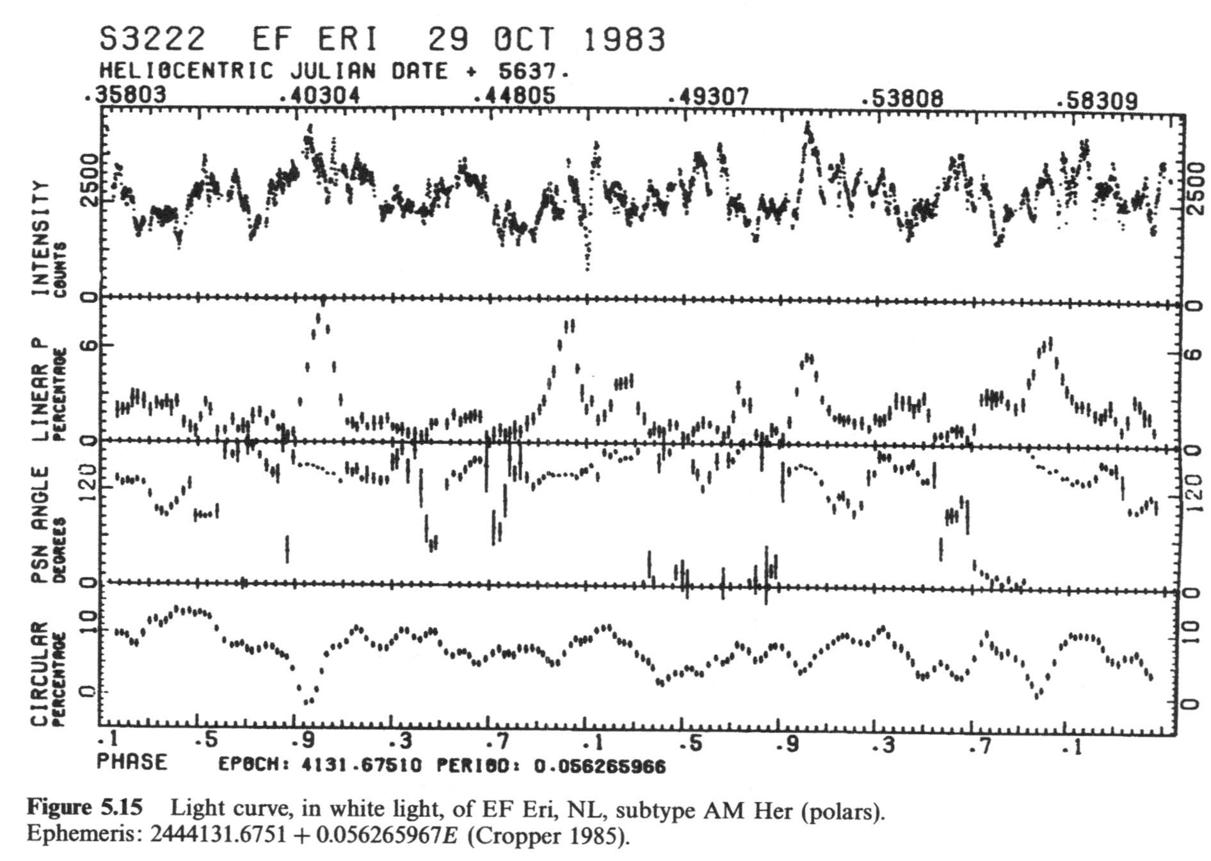

Figure 5.15 Light curve, in white light, of EF Eri, NL, subtype AM Her (polars). Ephemeris: 2444131.6751 + 0.056265967E (Cropper 1985).

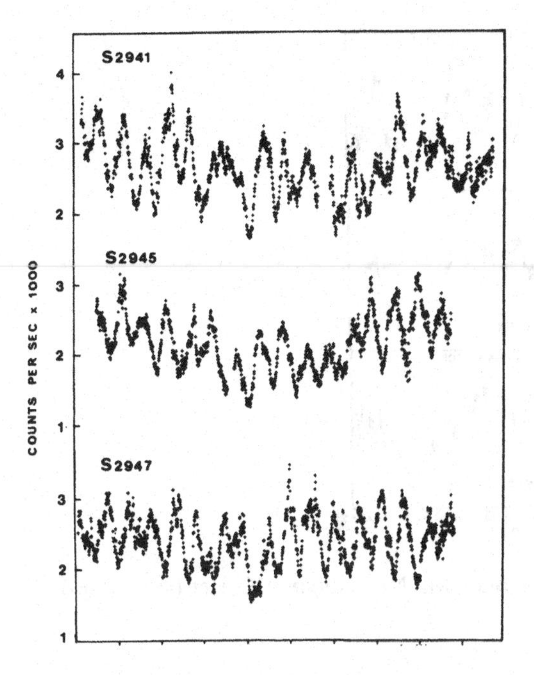

Figure 5.16 Light curve (white light) of BG CMi, NL, subtype DQ Her (= intermediate polar), ephemeris $2445020.3075 + 0.13480E$ (McHardy *et al.* 1984). For S 2941 the origin is at JD 2 445 020.2800; for S 2945 it is 2 445 022.3000 and for S 2947 it is 2 445 023.2800. The tick marks on the X-axis are $0^{\rm d}\!.02$ apart.

which is near the main sequence. Accretion occurs in the outer regions via a disc; close to the white dwarf the accretion disc is disrupted by the magnetic field and mass flows via an accretion column towards its magnetic poles. Light variations are caused by eclipse effects and by the rotationally modulated accretion effects (11 objects are known, among them a few objects that have shown nova explosions).

(iv) UX = UX UMa systems. These systems have bright accretion discs, caused by high accretion rates; often eclipse effects are observed in the light curves. Some systems resemble novae at minimum light (15 objects).

(v) VY = VY Scl stars (anti-dwarf novae). These systems are similar to UX UMa systems and most of the time they are in a 'high state'.

From time to time, fading by several magnitudes occurs, and the objects can remain for days to months in a 'low state' (11 objects). Most of them have orbital periods near 3 hours, and anti-dwarf novae are candidates for systems entering the 'period gap' (between 2 and 3 hours), in which cataclysmic variables are supposed to exist, but in a detached state: the secondary underfills

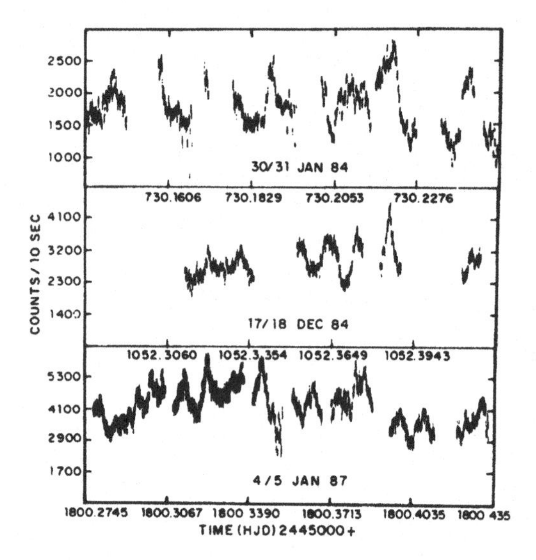

Figure 5.17 *B* light curve of BG CMi, NL, subtype DQ Her (= intermediate polar) (Vaidya *et al.* 1988).

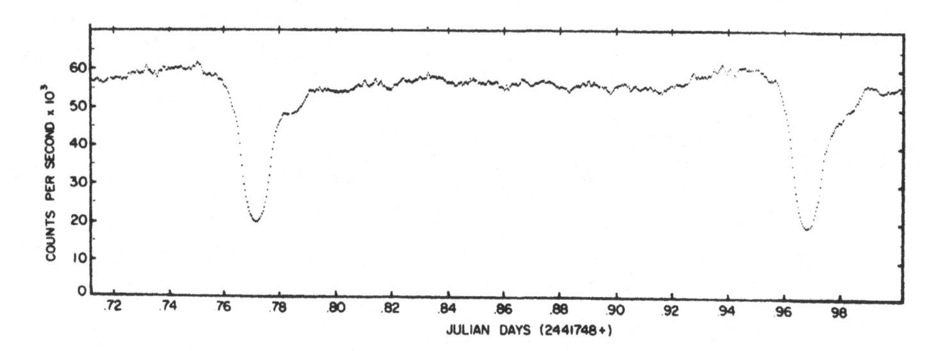

Figure 5.18 Light curve, in white light, of UX UMa, NL, subtype UX UMa, from Nather & Robinson (1974). Note the presence of eclipses of the disc (deep eclipse) and the 'hot spot' (subsequent shallow eclipse feature).

its critical Roche volume, mass transfer and outburst activity ceases (11 objects). Some NL cannot be fitted into one of these subclasses (6 objects).

Information on nova-like stars, dwarf novae and novae is to be found in Hack & La Dous (1993) and in Mauche (1990). Ritter (1990) compiled a catalogue of cataclysmic binaries, low-mass X-ray binaries and related objects, a catalogue and atlas of all known cataclysmic variables (dwarf novae, nova-like stars, and novae) was compiled by Downes & Shara (1993).

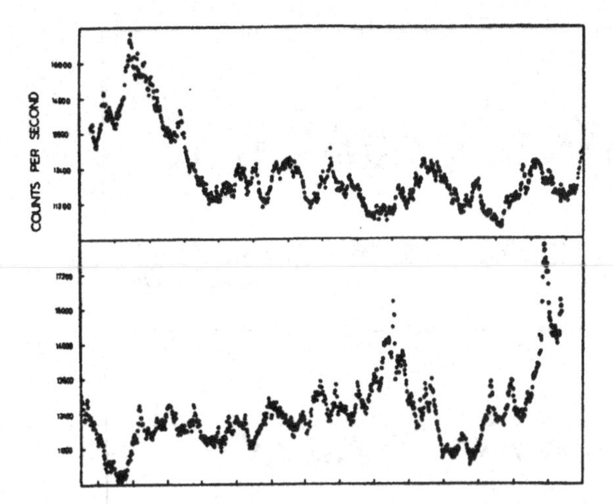

Figure 5.19 Light curve, in white light, of VY Scl, NL, subtype VY Scl (= anti-dwarf novae). The X-axis starts at 2 441 839.5655, the tick marks are at intervals of 0ᵈ.0035 (Warner & van Citters 1974).

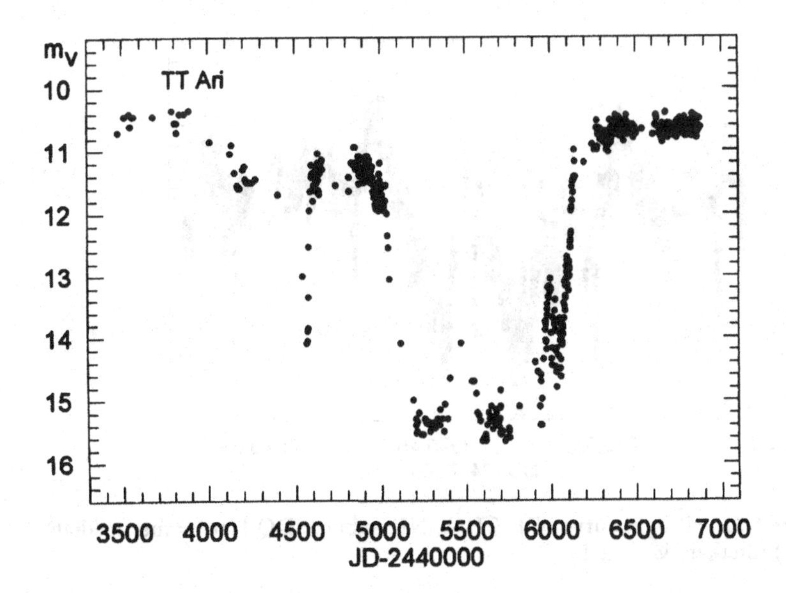

Figure 5.20 Visual light curve of TT Ari, NL, subtype VY Scl. Data from AAVSO.

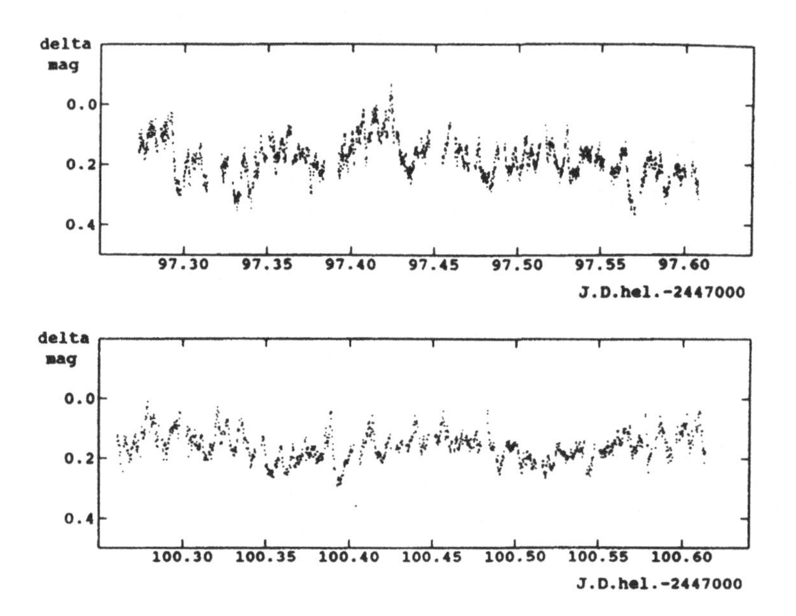

Figure 5.21 Differential *V* light curves of TT Ari, NL, subtype VY Scl (= anti-dwarf novae). The comparison star is BD +14 339 (Udalski 1988).

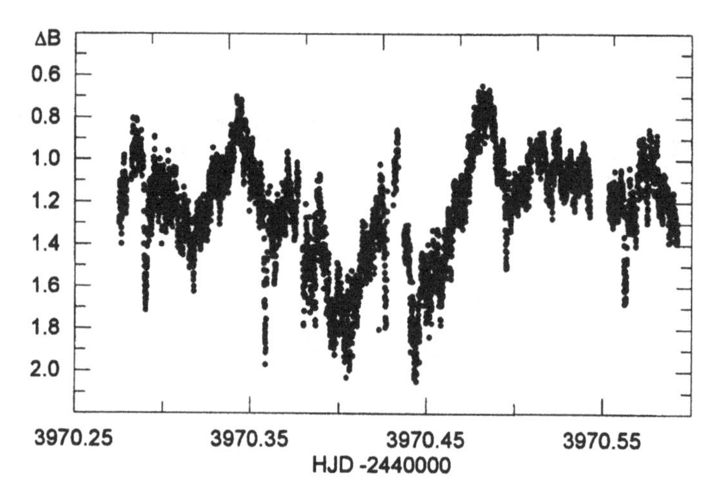

Figure 5.22 *B* light curve of EX Hya, NL, subtype DQ Her (= intermediate polar) (Sterken & Vogt 1995).

5.4 Dwarf novae

N. Vogt

The U-Geminorum stars or 'dwarf novae' (UG) form an important sub-group of the so-called 'cataclysmic variable stars', which are interacting binaries. They consist of a white-dwarf primary and a red main-sequence star (spectral type between G and M) as secondary. The companion is so close to the primary that it permanently loses matter from its surface towards the white dwarf. This stream, however, cannot directly hit the white dwarf, because its large angular momentum forces the expelled matter in keplerian orbits around the white dwarf. As such, an 'accretion disc' is formed. The light variations observed in many cataclysmic variables are caused by the rather complex behaviour of this accretion disc.

Dwarf novae are characterized by semi-regular 'outbursts': a sudden brightening by 3–8 magnitudes within about 1 day, a bright phase of 3–10 days and a subsequent decline which also lasts a few days. Outbursts repeat at mean intervals ('cycle length') characteristic for each star. The shortest 'cycles' are about 10 days. Typical cycle lengths range from 20–200 days, but exceptionally large ones up to 32 years are also present (see WZ Sge in Figs. 5.39, 5.40 and 5.41).

In their quiescent phases, dwarf novae have emission-line spectra with broad, often doubled Balmer and He I lines. These lines are formed in and around the accretion disc; they disappear during outburst when they are replaced by very broad Balmer absorption lines, which also originate in this disc which is, in these stages, much brighter and also optically thick.

Dwarf novae are – as all cataclysmic variables – close binaries with orbital periods between 80 minutes and 14 hours. It is well established now that their outbursts are due to the release of potential energy within the accretion disc: a sudden change of the viscosity in the disc enables the outer disc matter to 'fall' towards the white dwarf and release energy when hitting its surface ('accretion'). This energy is mainly transformed into radiation in the optical, ultraviolet and X-ray range. There is still some controversy about the reason for this change: while most people favour disc instabilities (e.g. change from radiative to convective disc structure: Meyer & Meyer-Hofmeister 1981, Ludwig *et al.* 1995), some defend the alternative view that the mass flow towards the disc is variable due to an (unknown) instability of the red secondary star (Bath *et al.* 1986).

The first dwarf nova detected was U Gem (Fig. 5.23 and 5.24), whose earliest-known outburst was observed by the English astronomer John Russel Hind (1823–1895) in December 1855. He classified U Gem as a faint nova. After the detection of another outburst in March 1856, it became clear that U Gem was a new type of variable star. The brightest dwarf nova is SS Cyg (Fig. 5.25), magnitude 8 at maximum and magnitude 12 at minimum. Dwarf novae are intrinsically faint and nearby objects ($M_v \approx +8$, distance ≥ 70 pc).

Amateur astronomers have contributed very much to the study of dwarf novae since professionals – who are subjected to tight scheduling of telescope

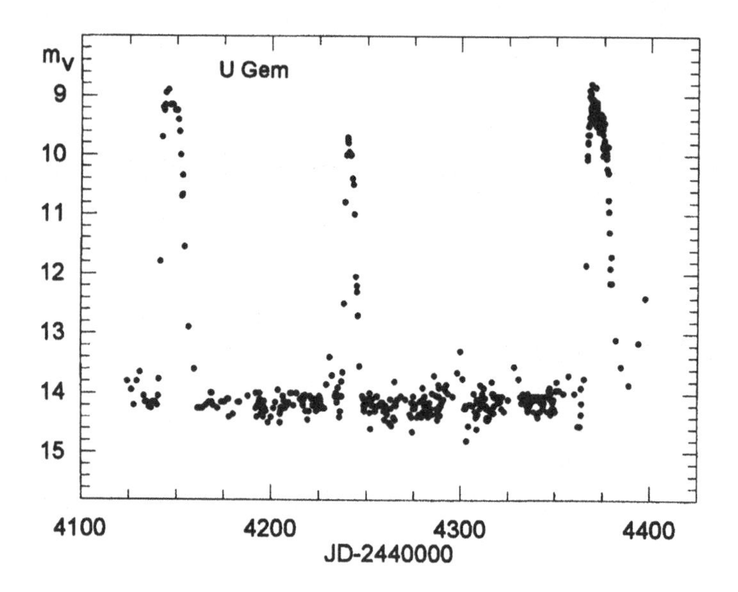

Figure 5.23 Visual light curve of U Gem (AAVSO).

time – are not able to follow the unpredictable outburst behaviour. Therefore, interesting facts have been extracted by analysing the extensive records of light curves obtained by amateurs. However, many questions are still open and amateurs will give important contributions to this area also in the future.

The study of these fascinating objects is also of more general impact in modern astrophysics: dwarf novae are ideal model cases to study the accretion physics which is very important in describing processes occurring in X-ray binaries, galactic nuclei, quasars and in the formation of stars. All these objects are less accessible and more difficult to observe than dwarf novae whose discs display different states of activity which can be studied in the same star. Therefore, dwarf novae are key objects for the study of accretion processes.

There are several subclasses of dwarf novae, viz., SS Cygni variables (UGSS), Z Camelopardalis variables (UGZ) and SU Ursae Majoris variables (UGSU). Their particular properties are described below.

For the most recent monographs on cataclysmic variables we refer to Hack & LaDous (1993) and Warner (1995), both with extended sections concerning dwarf novae. Physical parameters and recent references on those dwarf novae whose orbital periods are known are given by Ritter & Kolb (1995) while all known cataclysmic variables (including dwarf novae) are listed in Downes & Shara (1993).

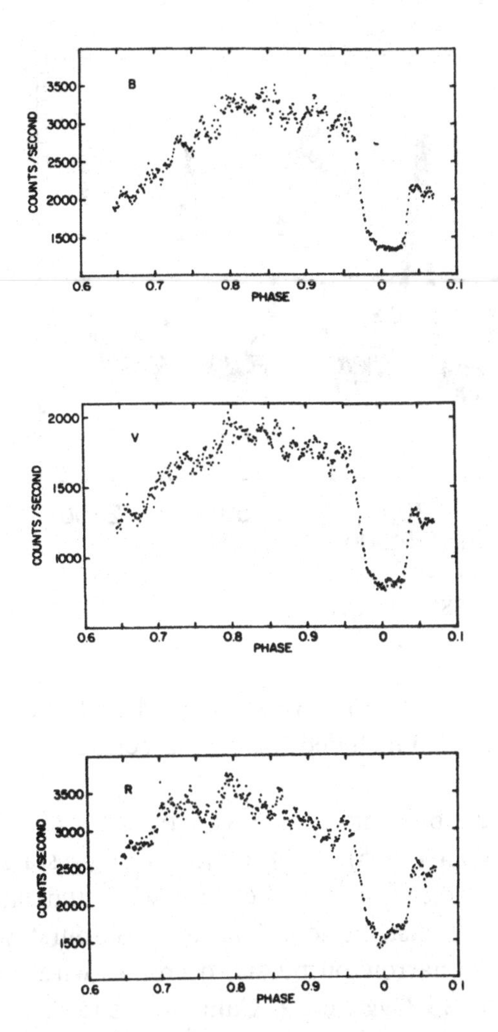

Figure 5.24 *B*, *V*, *R* light curve of U Gem in quiescent state according to ephemeris 2437638.82625 + 0.17690618*E* (mid-eclipse) (Zhang & Robinson 1987). The maximum (near phase 0.8) refers to the 'hump' which corresponds to the instant when the hot spot is directed towards the observer. At phase 0 the disc is partially eclipsed by the secondary.

SS Cygni variables

With 'SS Cygni stars' we denote all those dwarf novae which do not show peculiarities as the ones described under the Z Camelopardalis (UGZ) and SU Ursae Majoris stars (UGSU). The SS Cyg stars have orbital periods longer than 3 hours and show outbursts with typical intervals of 30–100 days, each lasting for about 3–10 days. For several decades it has been known that there is a relationship between outburst amplitude and cycle length – that is, the mean time interval between two subsequent outbursts, or the so-called Kukarkin-

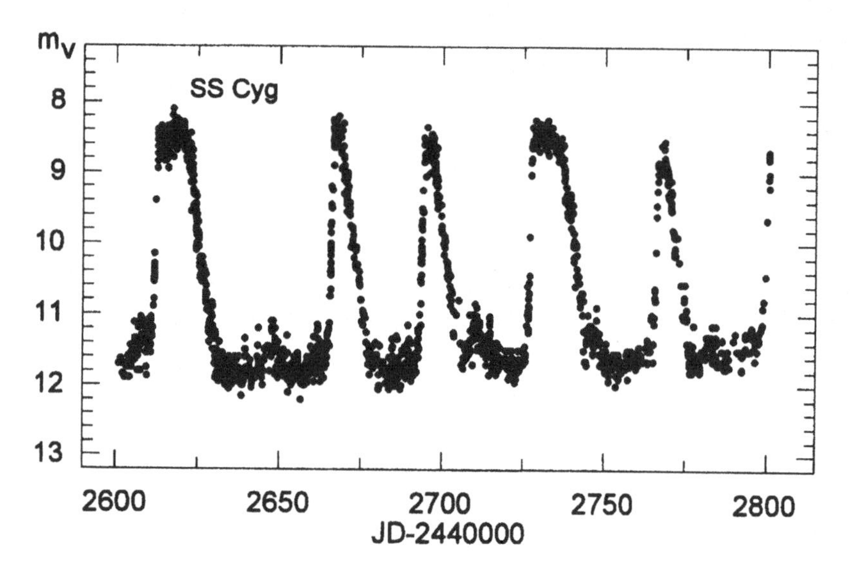

Figure 5.25 Visual light curve of SS Cyg (AAVSO).

Parenago relation. Richter & Braeuer (1989) have reanalysed and confirmed this relationship with new data, they also included the orbital period as a third, very important parameter.

Other peculiarities in the outburst behaviour, also as a function of orbital period, were recently studied by Vogt (1995): 75% of the SS Cyg stars (including some Z Cam stars) show a clear division between wide and narrow outbursts, revealing a bimodal width distribution. The total outburst width increases, the width ratio (of wide to narrow outbursts) decreases with orbital period. In addition, about 70% of SS Cyg and Z Cam stars show, at least occasionally, 'anomalous outbursts', characterized by an abnormally-slow rise phase which lasts nearly as long as the decline, and gives a symmetric appearance to the outburst light curve. For SS Cygni stars with long orbital periods, $P > 10^h$, anomalous outbursts seem to be the usual, if not exclusive, type of outbursts. Anomalous outbursts may be either wide or narrow. There is no obvious explanation for these peculiarities in the outburst behaviour.

Typical examples are

(i) U Gem: first dwarf nova detected; it is an eclipsing system with partial eclipses of the accretion disc (Figs. 5.23 and 5.24). For a recent discussion of its outburst behaviour, see Smak (1993).

(ii) SS Cyg: brightest and best observed of all dwarf novae, a most typical

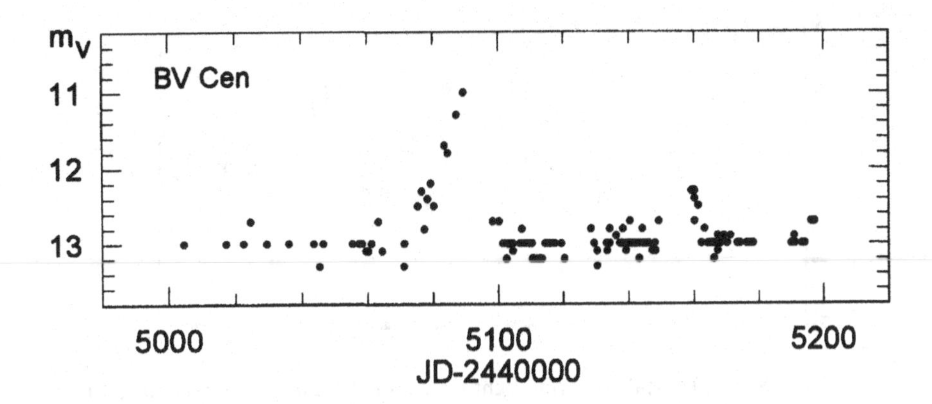

Figure 5.26 Visual light curve of BV Cen (AAVSO).

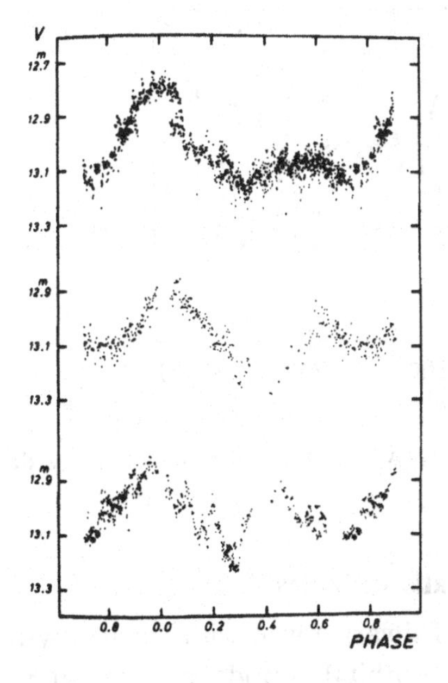

Figure 5.27 V hump light curve of BV Cen in quiescent state (ephemeris 24440264.78 + 0.6096144E, comparison star HD 117386, Vogt & Breysacher 1980). The hump maximum (here at phase 0) refers again to the hot spot conjunction, but an eclipse does not occur due to a lower orbital inclination of this system, compared to U Gem.

example (Fig. 5.25). For a detailed comparison between the observed behaviour and outburst models see Cannizzo (1993).

(iii) BV Cen: the dwarf nova with the longest orbital period $P = 14^{\mathrm{h}}\!.64$; it shows short 'standstills' in the rising branch of outbursts (Fig. 5.26). A typical hump light curve in quiescence is shown in Fig. 5.27 (Bateson 1974, Vogt & Breysacher 1980).

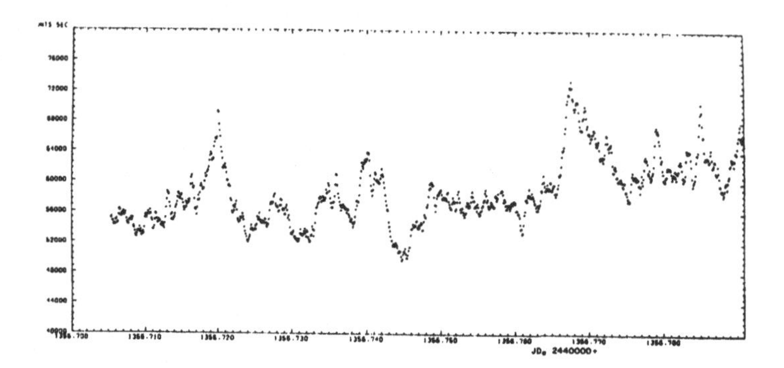

Figure 5.28 Typical flickering light curves of Z Cam (Y axis: counts s^{-1}) (Robinson 1973).

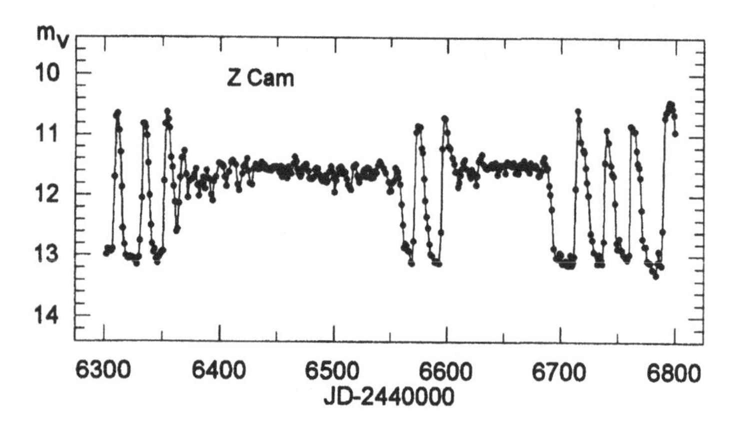

Figure 5.29 Visual light curve of Z Cam (AAVSO). For the sake of clarity, dots have been connected by a line.

Z Camelopardalis variables

Z Camelopardalis stars are a subclass of dwarf novae with frequently-repeating outbursts (every 10–30 days) and with orbital periods longer than 3 hours. At irregular intervals the star does not return to minimum brightness after an outburst: instead, it remains at an intermediate luminosity without major variation for several months or even years. These 'standstills', typical for Z Cam stars, are interpreted as a transitory stabilisation of the mass transfer and mass accretion rate at an intermediate value in dwarf novae with relatively large mass transfer rates (Meyer & Meyer-Hofmeister 1983). A standstill always ends with the return to quiescent state and the subsequent recovery of the outburst activity. Spectra in standstill are similar to those of dwarf novae in outburst (broad Balmer absorption).

There exist also extreme cases which are nearly always (for example, TT Ari) or always (for example, UX UMa) in standstill. Since these stars do not show

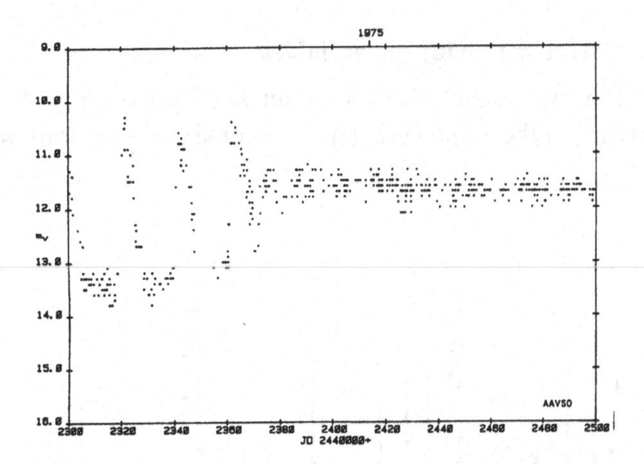

Figure 5.30 Portion of the AAVSO light curve of Z Cam, showing the progressive decrease of the outburst amplitude before entering standstill (Szkody & Mattei 1984).

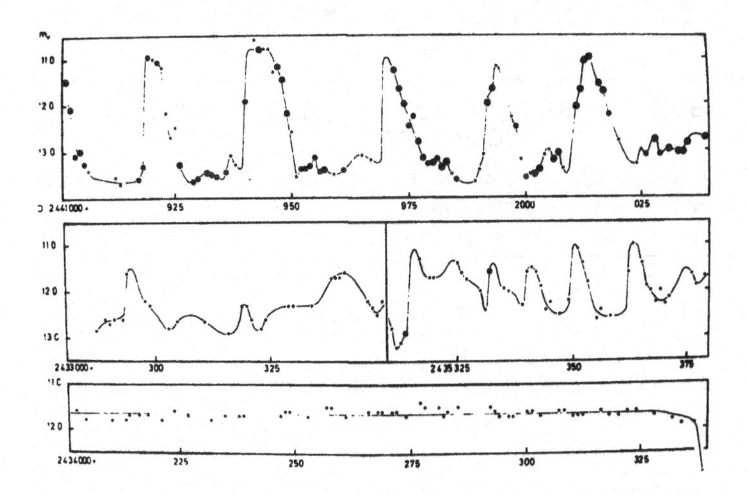

Figure 5.31 Visual light curves of RX And showing intervals of low (upper panel) and erratic (centre) activity and standstill. Increasing dot size indicates means of 1–2, 3–4 and >5 observations. The solid lines only suggest the possible course of variation (Howarth 1977).

typical dwarf nova outbursts, they normally are classified among the 'nova-like stars' (see Section 5.3).

Typical examples of Z Cam stars are Z Cam (Figs. 5.28, 5.29 and 5.30), the prototype of this subclass, and one of the brightest members, and RX And (Fig. 5.31).

SU Ursae Majoris variables

SU UMa stars are dwarf novae which are characterized by two very distinct types of outbursts occurring in the same star: (i) frequent short eruptions which

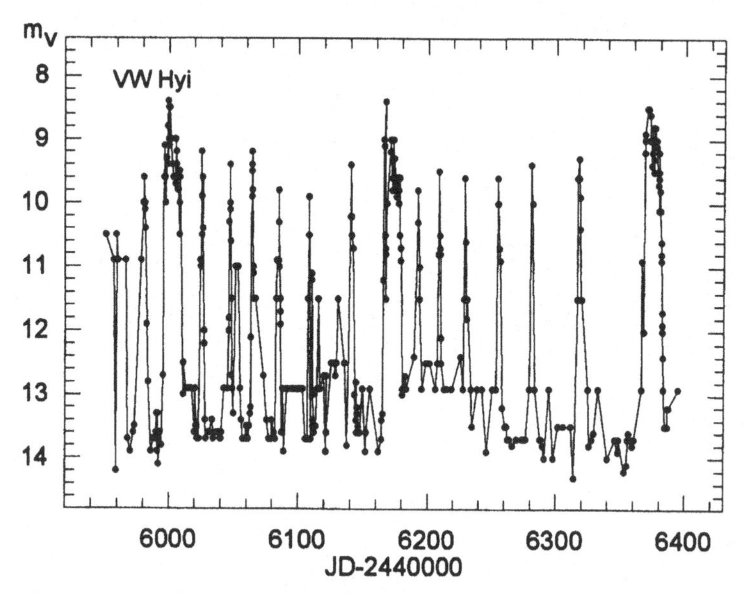

Figure 5.32 Visual light curve of VW Hyi (Bateson 1977); for the sake of clarity, the dots have been connected by a line.

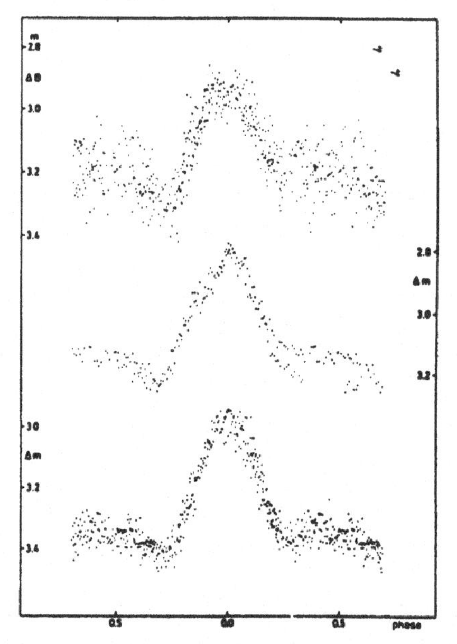

Figure 5.33 Orbital hump light curves (in white light) of VW Hyi in quiescence (Vogt 1974), at three different epochs (March 1972, January 1974 and March 1974).

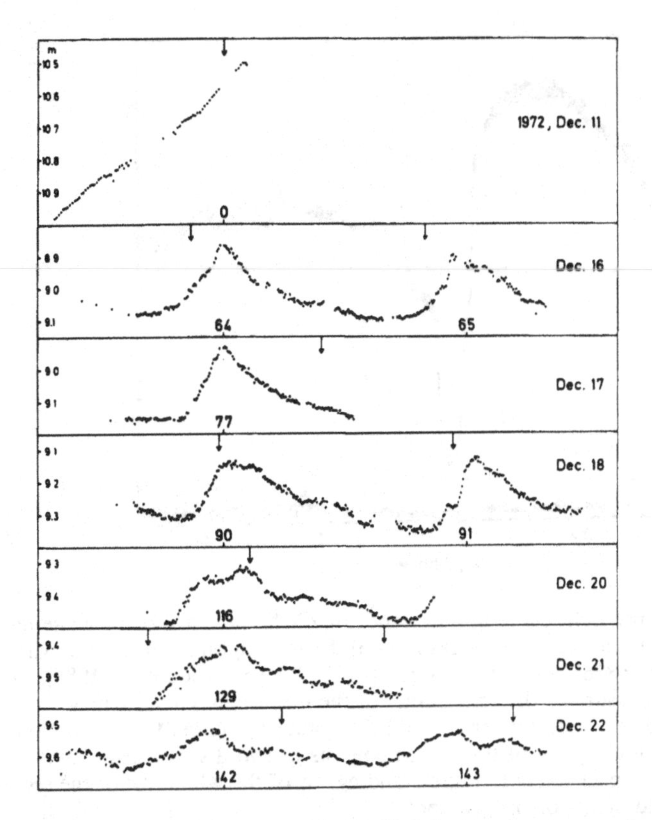

Figure 5.34 Magnitudes of VW Hyi during the superoutburst of December 1972. V magnitudes, except for cycles 0 and 143, where only B was observed. The X-axis gives superhump cycle numbers according to $2441662.687 + 0.07676E$, the arrows correspond to phase 0 according to the quiescent-hump ephemeris $2440128.0223 + 0.0742711E$ (Vogt 1974).

repeat at intervals of 15–40 days and last only a few days; and (ii) less-frequent long eruptions or 'superoutbursts' which last 10–20 days and occur at intervals of six months to several years. Superoutbursts are also brighter than short eruptions, by about 1 magnitude. Shortly after having passed maximum the star develops the so-called 'superhumps', periodic peak structures in the light curve, which repeat with a period P_s that is 3–5% longer than the orbital period P_o. The occurrence of superhumps is not limited to binary orbits with high inclination (as the orbital hump): superhumps are observed at all angles of inclination. In general, SU UMa stars have short orbital periods which are either below the period gap ($P_o \lesssim 2$ h) or, in a few cases, near its upper border ($P_o \approx 3$ h).

First hints towards a physical understanding of this peculiar superhump behaviour came from spectroscopic observations of Z Cha by Vogt (1982)

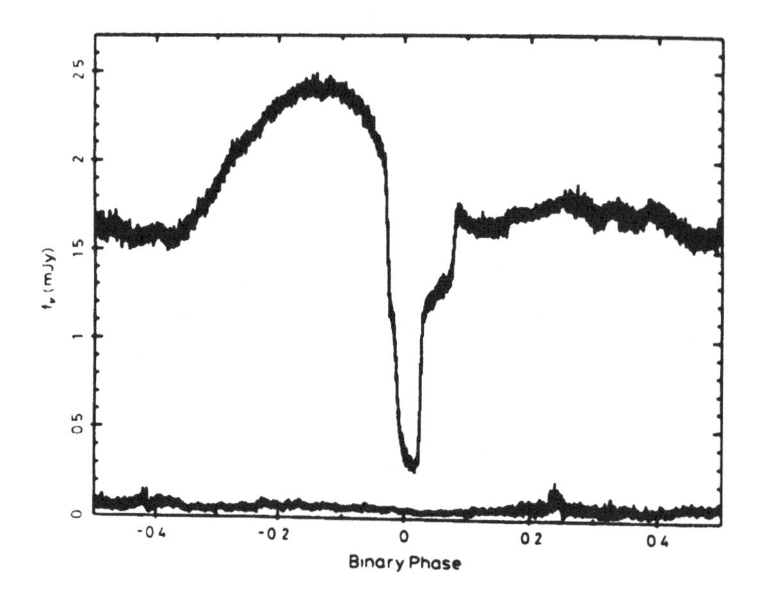

Figure 5.35 Eclipse light curve in quiescence of Z Cha (average of 17 white-light curves observed at one second time-resolution). Phase diagram constructed with ephemeris $2440264.68261 + 0.0744992335E + 6.1710^{-13}E^2$ (Wood *et al.* 1986). The lower curve is a measure of the uncertainty in the mean light curve. 1 mJy corresponds to 3300 counts s^{-1} for a $V = 13\overset{m}{.}75$ star at a 75 cm telescope. Note that ingress as well as egress of the eclipse consists of two distinct steep parts: in both cases, the first part refers to ingress and egress of the white dwarf, the second part to ingress and egress of the hot spot.

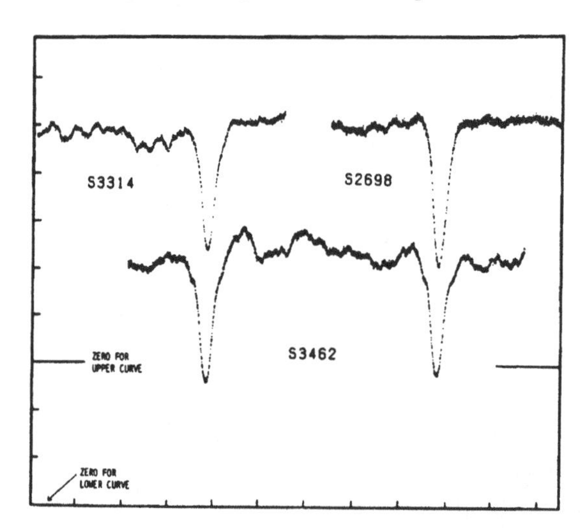

Figure 5.36 Light curves during supermaxima, but before the start of superhumps (S3314 and S3462) and during an ordinary outburst of Z Cha (S2698). The scale of the Y-axis is normalised to unity for regions of the light curve free of effects of eclipse or humps. The tic marks on the X-axis are $0\overset{d}{.}015$ apart (Warner & O'Donoghue 1988).

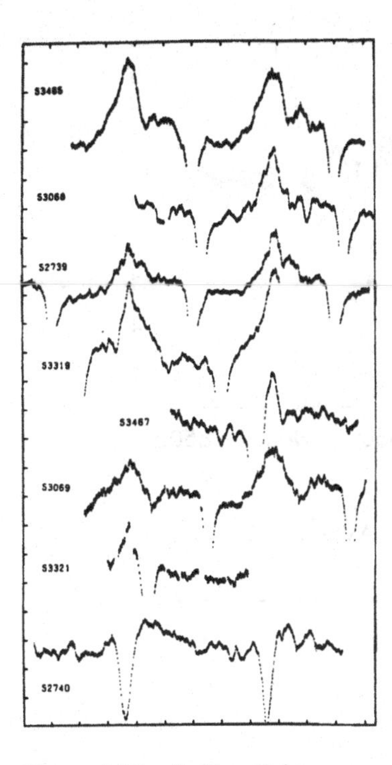

Figure 5.37 Z Cha: light curves during supermaxima, aligned according to superhump phase and displayed in order of time after beginning of outburst. The light curves are scaled vertically so that regions free of interference from the superhump or eclipse are normalised to unit intensity. The light curves are spaced vertically by 0.5 of these units. The tic marks on the X-axis are $0^{\mathrm{d}}.015$ apart (Warner & O'Donoghue 1988).

during superoutburst: the orbital disc motion revealed rather large displacements (several 100 km s^{-1}) in the zero term of the radial velocity curve which turned out to vary with the beat period between orbital and superhump periods. This must be taken as a direct evidence for an elliptical deformation of the accretion disc during superoutburst. The disc is slowly precessing in the inertial frame of reference. Subsequent theoretical calculations as well as hydrodynamic 2-D simulations showed that the eccentric disc is a natural consequence of its expansion towards the critical radius of the 3:1 resonance: a precessing eccentric disc develops due to a tidally-driven instability. Periodic tidal stressing of the eccentric disc by the orbiting secondary gives rise to the superhump light variations (Whitehurst 1988, Ichikawa *et al.* 1993, and references therein). The tidal instability implies binary configurations with extreme mass ratios ($m_1/m_2 \geq 4$) for SU UMa stars.

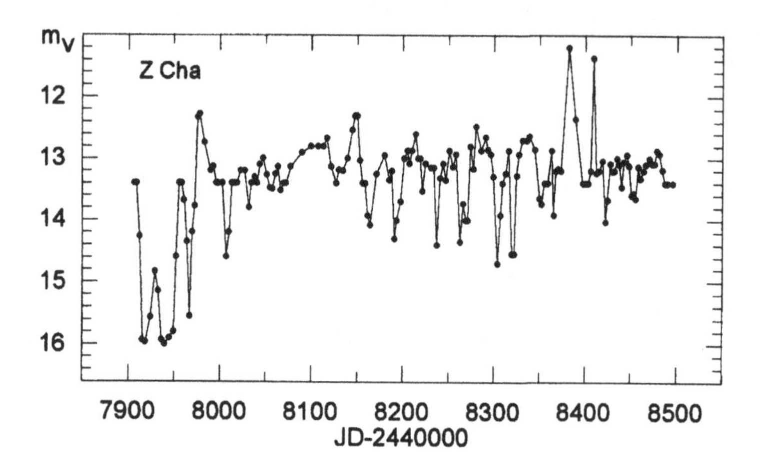

Figure 5.38 Visual light curve of Z Cha (AAVSO).

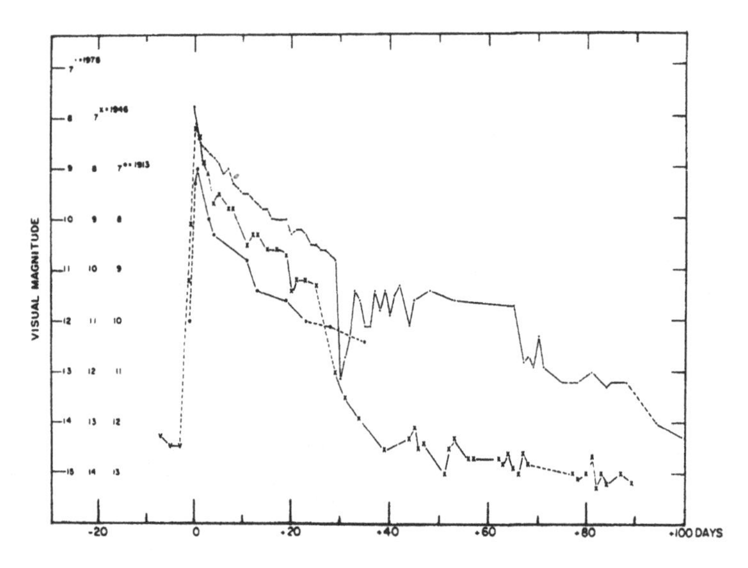

Figure 5.39 Visual light curve of WZ Sge (AAVSO). The X-axis is in days after the 1913, 1946 and 1978 outbursts.

For a recent compilation of all known SU UMa stars and their most important observational parameters see Molnar & Kobulnicky (1992).

Typical examples are

(i) VW Hyi, the brightest and best studied SU UMa star that shows an 'orbital hump' in the light curve at quiescence (Figs. 5.32, 5.33 and 5.34).

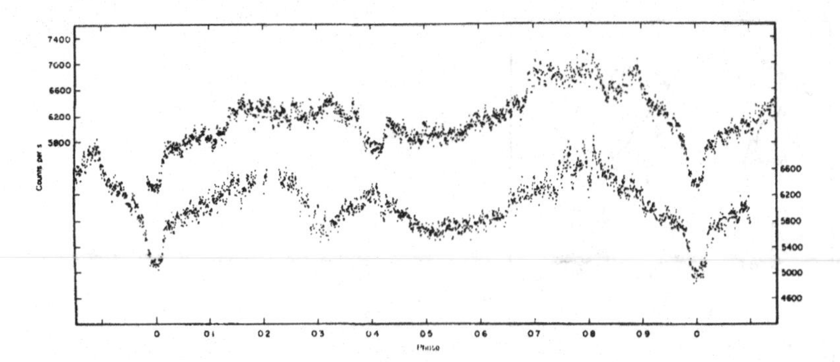

Figure 5.40 Quiescent white-light phase diagram in counts s^{-1} of WZ Sge, ephemeris $2437547.72845 + 0.0566877847E$ (Warner & Nather 1972b). The lower curve is a continuation of the upper curve, and includes some overlap.

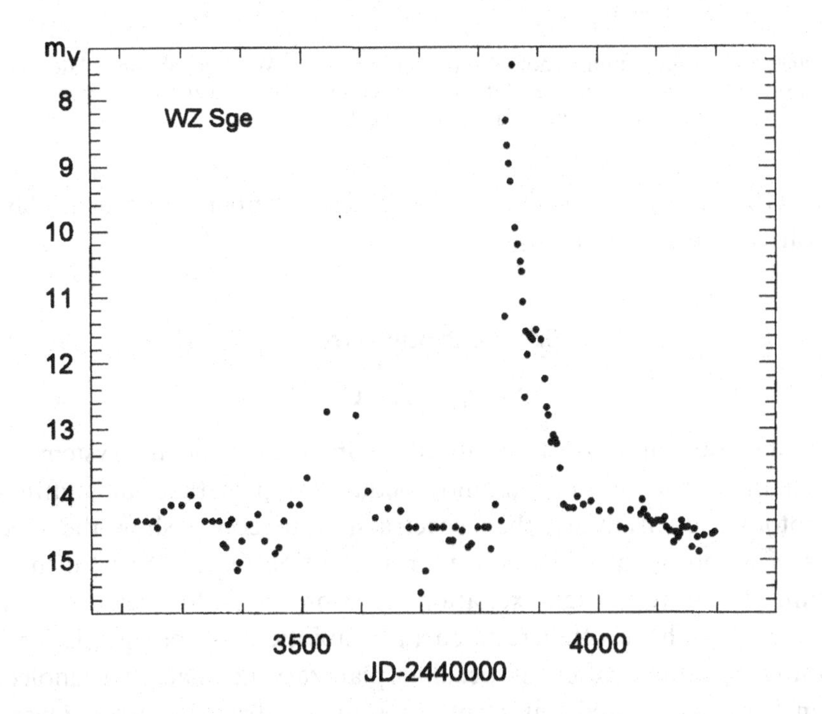

Figure 5.41 Visual light curve of WZ Sge (AAVSO). Detail of the 1978 outburst.

(ii) Z Cha, an SU UMa star with total eclipses of white dwarf and hot spot (Figs. 5.35–5.38)

(iii) WZ Sge, a dwarf nova with the longest-known outburst cycle (32 years), at the same time it has the shortest-known orbital period ($1^h\!.36$), Figs. 5.39–5.42. WZ Sge is an eclipsing system, an extreme case of the

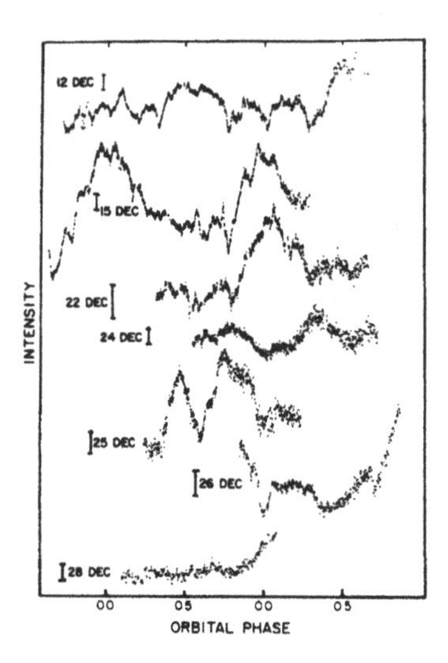

Figure 5.42 Superhump photoelectric light curves of WZ Sge, obtained late in the 1978 eruption, with a time resolution of 5 seconds. The vertical bars represent a $0^{\mathrm{m}}.05$ variation on each night (Patterson *et al.* 1981).

SU UMa class, or possibly the brightest member of its own class of dwarf novae (Osaki 1995).

5.5 Symbiotic stars

P.A. Whitelock

The symbiotic stars or Z And variables are interacting binary systems. The defining characteristic of this inhomogeneous group is that, in addition to erratic photometric variability, their spectra simultaneously show the spectral signatures of a cool giant (e.g., molecular absorption features) and of a high-temperature plasma (e.g., high-excitation emission lines). Studies over limited wavelength regions have often resulted in symbiotic stars being misclassified as something else, most often as peculiar planetary nebulae. The molecular absorption features are frequently only present in infrared spectra. Over 150 stars have been classified as symbiotic.

The giant component of the binary system is usually of spectral type M or C. There are also a few so-called 'yellow symbiotics' which have G-type spectra, but the relationship of these objects to the rest of the class is uncertain. The other star in the binary may be a low-mass main-sequence star or compact object, e.g., subdwarf, white dwarf or neutron star. The interaction which results in what is often described as the 'symbiotic phenomenon' (including erratic variability and high-excitation emission lines) arises when mass is transferred

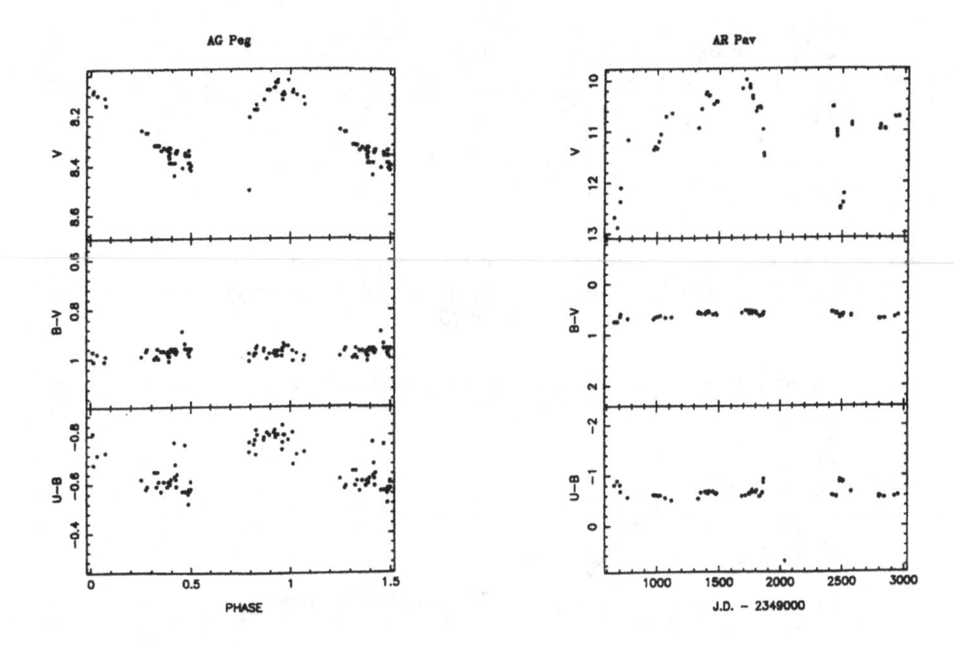

Figure 5.43 $V, B-V, U-B$ light curve of AG Peg (data from Belyakina 1970), ephemeris 2442710.1 + 816.5E (from Fernie 1985), the comparison star was BD +11 4681 and of AR Pav (data and ephemeris 2420330.0 + 604.6E from Andrews 1974).

from the giant to the other star. Most of the well studied systems contain either a main sequence star which accretes by direct tidal overflow (Roche-lobe overflow) from the giant, or a white dwarf which accretes from the giant's stellar wind. Many systems show evidence of an accretion disc. The process of mass transfer often gives rise to a 'hot spot'. In many cases this 'hot spot' provides the temperature necessary for ionizing part of the circumstellar environment and producing the emission lines. The symbiotic stars are closely related to the even rarer VV Cep systems within which a late-type supergiant interacts with an O or a B star.

The level of symbiotic activity must depend critically on the separation and evolutionary states of the two stars. The time scale of the variability ranges from seconds (i.e., 'flickering') to many decades (i.e., apparently secular) although the best studied variations in the well studied systems are usually on a time scale of days to tens of days. Most systems show variability on a variety of time scales. Large-amplitude nova-like variability also occurs and is discussed below. Variations due to orbital modulation are sometimes seen, as the light curve for AG Peg in Fig. 5.43 illustrates. Orbital periods vary from

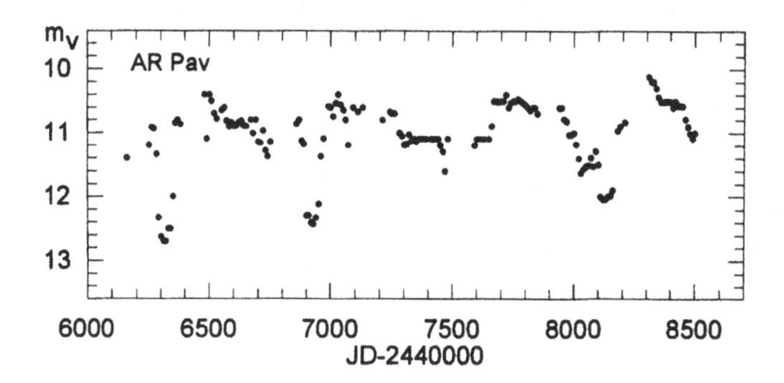

Figure 5.44 Visual light curve of AR Pav (AAVSO).

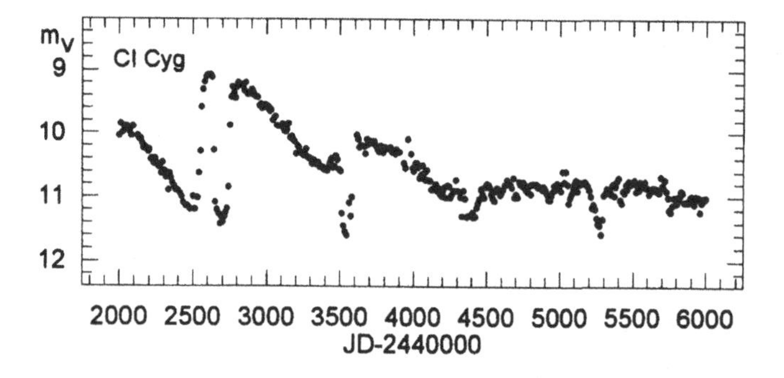

Figure 5.45 Visual light curve of CI Cyg (AAVSO).

around 100 days to, probably, many years. Some of the better studied systems are known to be eclipsing, e.g., AR Pav (Fig. 5.44) and CI Cyg (Fig. 5.45).

The 'very slow novae' form an interesting subgroup of the symbiotic stars. These include RR Tel (Fig. 5.46), V 1016 Cyg and AG Peg. Such objects have undergone outbursts of several magnitudes which decayed on a time scale of years to decades. It is generally thought that these outbursts are caused by a thermonuclear runaway in accreted material on the surface of a white dwarf. They are therefore closely related to the recurrent novae, some of which have M giant components and are often classified with the symbiotic stars, e.g., RS Oph and T CrB.

The Mira or D-type symbiotics are an important and relatively homogeneous sub-group distinguished by their characteristic near-infrared colours which indicate the presence of dust (as opposed to S-type symbiotics which have stellar infrared colours). Near-infrared studies suggest that the Miras in these systems are normal, though subject to higher circumstellar extinction than is usual for single stars of this type. In most cases the Mira-like variability is only

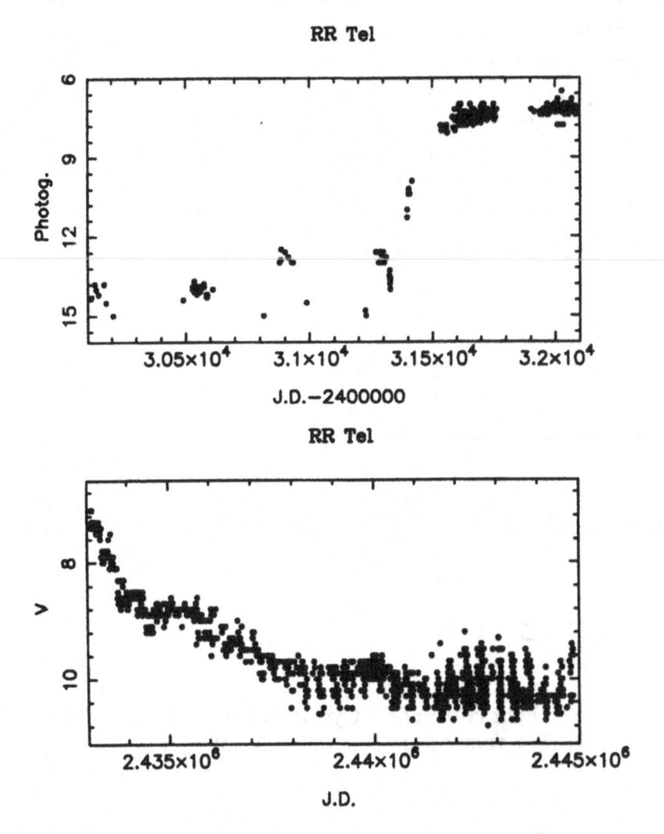

Figure 5.46 Photographic and visual light curve of RR Tel. The photographic magnitudes are from Mayall (1949), the visual estimates from Bateson & Kenyon (1991, RASNZ private communication).

clearly marked in the infrared while the visual light is dominated by the activity of the hot component. R Aqr is the most notable exception to this, as it shows clear Mira-like variability in the visual. The Mira symbiotics probably have the longest orbital periods of all known interacting binaries; no periods have been determined with certainty although 44 years has been suggested for R Aqr. Mira itself (*o* Ceti) forms a binary system with its white-dwarf companion and probably only avoids being classed as symbiotic by the relatively wide separation of the two stars which results in only a very low level of interaction.

Symbiotic stars have been the subjects of limited photoelectric photometry although the situation has improved somewhat in recent years. Because the blue spectra of these stars are usually dominated by emission lines it is not possible to put photoelectric photometry of them onto a standard system, such as the *UBV* or *uvby* system. For this reason it is often unhelpful to combine observations made with different photometers, and only large numbers of measurements made with a single instrument will help to define the detailed behaviour of

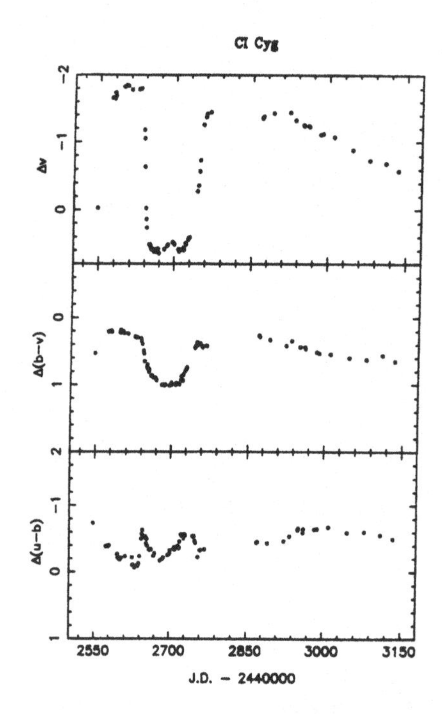

Figure 5.47 $V, b-v, u-b$ light curve of CI Cyg, data from Belyakina (1970), the period is 855 days. (1985). The comparison star was BD +35 3821

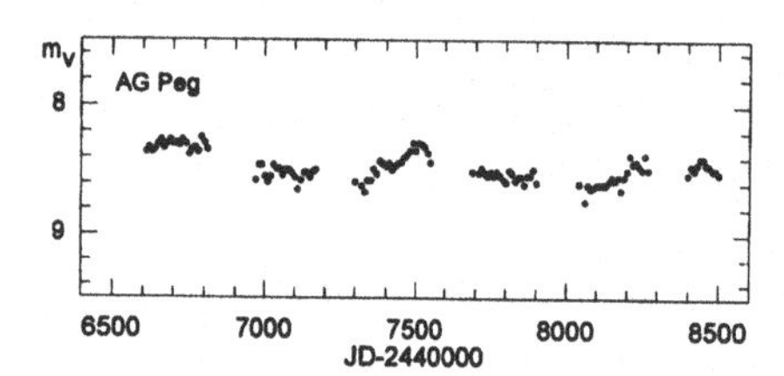

Figure 5.48 Visual light curve of AG Peg (AAVSO).

these stars. The visual observations of amateurs have proved particularly useful for following long-term trends and in alerting the community when a particular star enters an active phase.

The stars selected as representatives are:

(i) CI Cyg: an eclipsing system with an 812 day orbital period, comprising a Roche-lobe filling M4 giant and a main sequence accretor (Fig. 5.47). It has undergone several outbursts, no two of which seem to be identical. There is also some suggestion that it shows short time scale flickering.

(ii) AR Pav: An eclipsing system with a 605 day orbital period comprising a Roche-lobe filling M3 giant and a main sequence accretor (Fig. 5.43).

(iii) AG Peg: A bright (V ~ 8.4) well studied system comprising an M3 giant and probably a white dwarf (Fig. 5.48). The orbital period is $816^{d}.5$. It underwent a nova-like outburst in the 1850s from which it still appears to be declining.

(iv) RR Tel: A symbiotic Mira and very slow nova. Figure 5.46 shows the outburst as recorded photographically by Mayall (1949) and the post-outburst decay as recorded by the variable star section of the RASNZ. The data were made available by Scott Kenyon.

6
Eclipsing binary systems

6.1 Algol type eclipsing binaries

D. S. Hall

The Algol type eclipsing variables (EA) are a subgroup of the eclipsing binaries segregated according to light curve shape. The light remains rather constant between the eclipses, i.e., variability due to the ellipticity effect and/or the reflection effect is relatively insignificant. Consequently, the moments of the beginning and the end of the eclipses can be determined from the light curve. Eclipses can range from very shallow ($0^{m}_{.}01$) if partial, to very deep (several magnitudes) if total. The two eclipses can be comparable in depth or can be unequal. In a few cases the secondary eclipse is too shallow to be measurable (when one star is very cool), or absent altogether (highly eccentric orbit).

Light curves of this shape are produced by an eclipsing binary in which both components are nearly spherical, or only slightly ellipsoidal in shape. Though not explained in the *GCVS*, one component may be highly distorted, even filling its Roche lobe, provided it contributes relatively little to the system's total light. This is, in fact, the case for at least half of the known EA variables.

Among the EAs one may find binaries of very different evolutionary status:

(i) binaries containing two main-sequence stars of any spectral type from O to M, with CM Lac an example
(ii) binaries in which one or both components are evolved but have not yet overflowed their Roche lobes, with AR Lac an example
(iii) binaries with one star unevolved and the other overflowing its Roche lobe and causing mass transfer, with RZ Cas an example
(iv) binaries with one star highly evolved (a hot subdwarf or a white dwarf) and the other less evolved, like V 1379 Aql, or
(v) not evolved at all, like V 471 Tau.

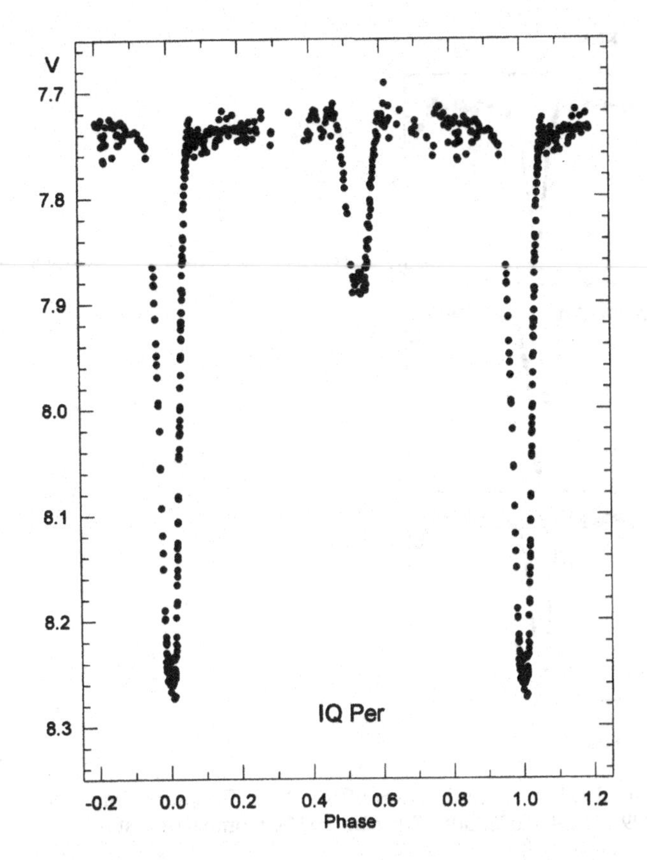

Figure 6.1 *V* phase diagram of IQ Per (EA) according to ephemeris 2440222.5974 + 1.7435673*E* (based on Tables III, IV and V of Hall *et al.* 1970).

Binaries in that third evolutionary state (semi-detached, one star evolved and one not, mass transfer in progress) are termed 'Algol-type binaries' or 'Algol-like binaries'. Such binaries, if eclipsing, can have light curves of the EA or EB shape, or they may not eclipse at all. Ironically, β Lyrae, the prototype of the EB light curve shape, is an Algol-type binary.

The first EA discovered, and the prototype of the group, was Algol = β Persei. Although its variability was known by the Chinese 2000 years before, John Goodricke in 1783 was the first to determine the strict periodicity of its variability ($2\overset{d}{.}867$) and the first to propose eclipses as the mechanism. Algol has partial eclipses, is semi-detached, undergoes mass transfer, has a chromospherically active secondary star which emits radio waves and X-rays, and belongs to a triple system.

Orbital periods range from extremely short (a fraction of a day) to very long (27 years for ϵ Aurigae). For an EA light curve shape, the stellar radius or radii

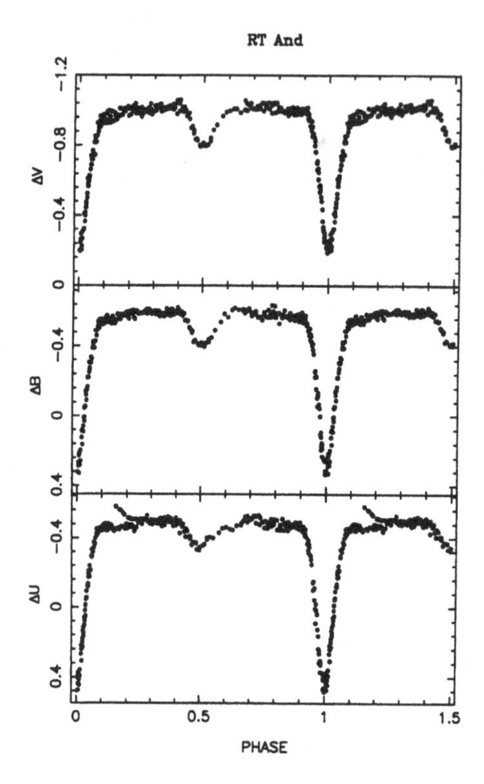

Figure 6.2 Differential *UBV* phase diagram of RT And (EA) according to ephemeris 2437999.7552 + 0.62893066*E* (Dean 1974). Comparison star: BD +52 3382.

must be a relatively small fraction of the star-to-star separation. Note that the brighter component in ϵ Aurigae is a *supergiant* but its radius is still a small fraction of the large (6000 solar radii) semi-major axis.

Orbital periods of EA's can be determined very accurately by timing the sharp eclipses. Period variations are found in many systems. Physical mechanisms responsible can be apsidal motion, orbit around a third body, mass loss and/or mass transfer, and solar-type magnetic cycles (Hall 1990a). The orbital period of Algol itself undergoes a 1.783 year cycle as it orbits around Algol C and it also has a 32-year magnetic cycle (Soderhjelm 1980).

The eclipsing binaries selected to illustrate light curves of the EA type are:

(i) IQ Per: a B7 V + A2 V 1.74-day system composed of two unevolved main-sequence stars. Therefore, primary eclipse is annular, while secondary eclipse is total (Fig. 6.1). A moderate orbital eccentricity displaces secondary eclipse away from phase 0.5. The period has been constant.

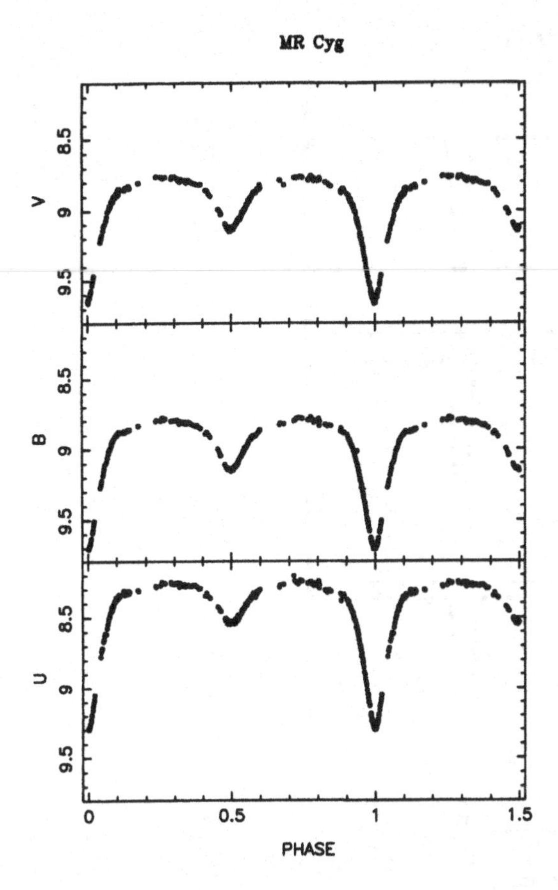

Figure 6.3 *UBV* phase diagram of MR Cyg (EA) according to ephemeris 2427013.612 + 1.677035*E* (Hall & Hardie 1969).

(ii) RT And: an F8 V + G5 V 0.63-day system (Fig. 6.2). At least one of the two stars is chromospherically active and produces a starspot wave in the light curve, so the system is also classified 'RS'. The orbital period is decreasing monotonically.

(iii) MR Cyg: a B3 V + B8 V 1.677-day system which may be semi-detached or nearly so. The ellipticity effect and the reflection effect are clearly noticeable (Fig. 6.3), but not extremely large. The orbital period is constant.

(iv) VW Cyg: an A3e + K1 IV 8.43-day system which is semi-detached, even though the *GCVS* classifies it as detached. Primary eclipse is total and very deep, but appears slightly rounded (see Fig. 6.4) during totality because of photometric effects due to circumstellar material. The period has, at times, decreased and increased.

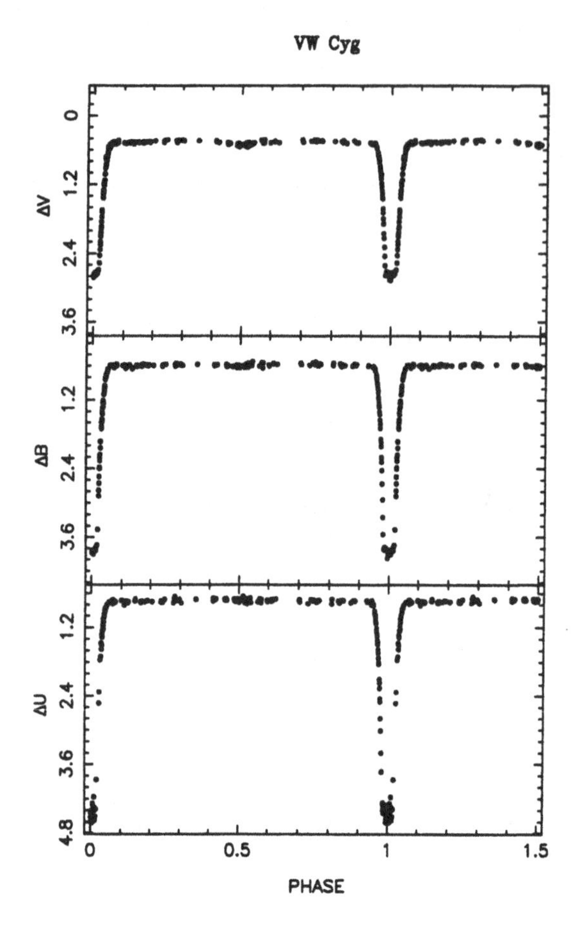

Figure 6.4 Differential *UBV* phase diagram of VW Cyg (EA) according to ephemeris 2441116.8678 + 8.4303102*E* (Ammann *et al.* 1979). Comparison star: BD +34 3945.

(v) RS Cep: an A5 IIIe + G8 IV-III 12.42-day system which is semi-detached, even though the *GCVS* classifies it detached. Primary eclipse is total and quite deep but appears slightly rounded during totality because of photometric effects of circumstellar material (Fig. 6.5). The period has increased.

The interested reader may find more details on these objects in Thomas (1977), Sahade & Wood (1978), Soderhjelm (1980), Olson (1985), Batten (1989), and Hall (1990a).

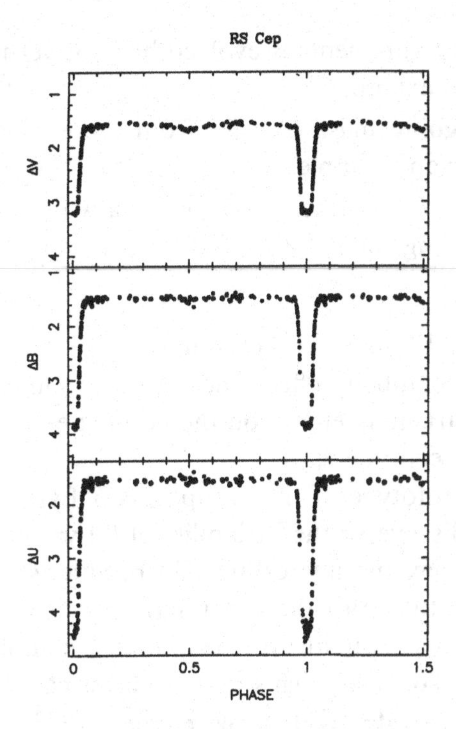

Figure 6.5 Differential *UBV* phase diagram of RS Cep (EA) according to ephemeris 2440862.677 + 12.420105*E* (Hall *et al.* 1984). Comparison star: BD +80 159.

6.2 β Lyrae type eclipsing binaries

D. S. Hall

The β Lyrae type eclipsing variables (EB) are another subgroup of the eclipsing binaries segregated according to light curve shape. The light curve varies continuously between eclipses, making it difficult to specify the moments of the beginning and the end of the eclipses. To distinguish between the EB's and the EW's, according to the *GCVS*, the former generally have primary and secondary eclipses significantly different in depth, orbital periods longer than a day, and spectral types B or A.

Light curves of this shape are supposed to be produced by an eclipsing binary in which one or both components is highly ellipsoidal. One of the components may even fill its Roche lobe.

Among the EB's one may find binaries of very different evolutionary status:

(i) unevolved binaries consisting of two main-sequence stars but a relatively short orbital period, with XY UMa an example

(ii) binaries in which one or both components is evolved but not yet filling the Roche lobe, with ζ And an example
(iii) semi-detached binaries undergoing mass transfer from the evolved to the unevolved star, with β Lyr an example
(iv) binaries with one star highly evolved (a hot subdwarf or a white dwarf) and the other, evolved, producing the ellipticity effect, with AP Psc an example.

Ironically, some binaries classified EB are not eclipsing at all. The light variation is produced entirely by the ellipticity effect and the two minima are unequal as a result of greater limb-darkening effects on the pointed end of the highly distorted star, as discussed in Section 4.2.

The first EB discovered, and the prototype of the group, was β Lyrae. The same John Goodricke of Algol fame discovered the variability of β Lyr one year later, in 1784. β Lyr is extremely complex and interesting. The brighter star fills its Roche lobe and is transferring matter onto the other star so rapidly that a thick (optically and geometrically) disc is built up which almost completely obscures the underlying mass-gaining star itself. This mass transfer causes the orbital period to increase at a furious rate (Klimek & Kreiner 1973, 1975; Bahyl *et al.* 1979). In the 210 years since Goodricke's 1784 timing, the period has increased from $12^{\mathrm{d}}.8925$ to $12^{\mathrm{d}}.93854$, an increase of 0.35%!

The binaries selected to illustrate the EB-type light curve shape are the following: The interested reader may find more details on these objects in Sahade & Wood (1978) and Plavec (1983, 1985).

(i) β Lyr: a B8 II + F system with an orbital period of 12.94 days as of 1990. At the deeper minimum the hotter B8 star is partly eclipsed by the accretion disc (Fig. 6.6). The F spectral type refers to the average surface temperature of the edge of the disc whereas the underlying star is about 10 solar masses and would appear much hotter, early B spectral type, if not obscured.
(ii) DO Cassiopeiae: an A2 V + G V 0.685-day system which the *GCVS* considers a contact binary (Fig. 6.7).
(iii) ζ Andromedae: a single-lined binary of spectral type K1 III, with an orbital period of $17^{\mathrm{d}}.77$ (Fig. 6.8). The K giant nearly, but not quite, fills its Roche lobe. Because the K giant is chromospherically active, the *GCVS* added the 'RS' classification, but only recently has it been shown (Kaye *et al.* 1995) that starspots contribute a little to the variability. Moreover, there is no evidence that eclipses are occurring at all. It is primarily an ellipsoidal variable.

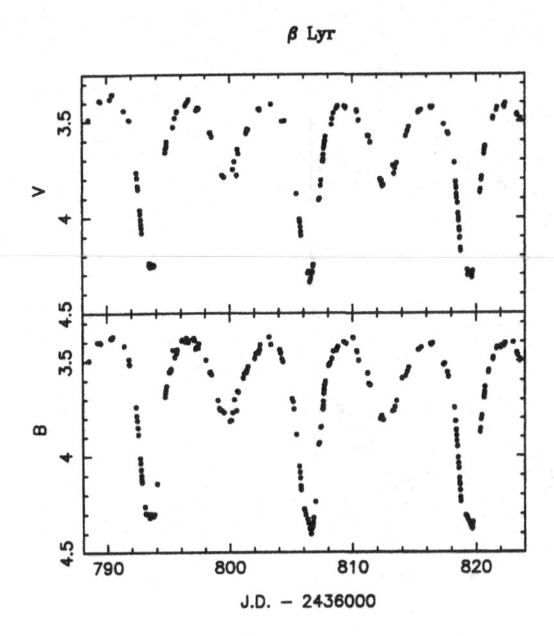

Figure 6.6 B, V phase diagram of β Lyr (EB) according to ephemeris 2436806.405 + 12.9355E (Larsson-Leander 1969).

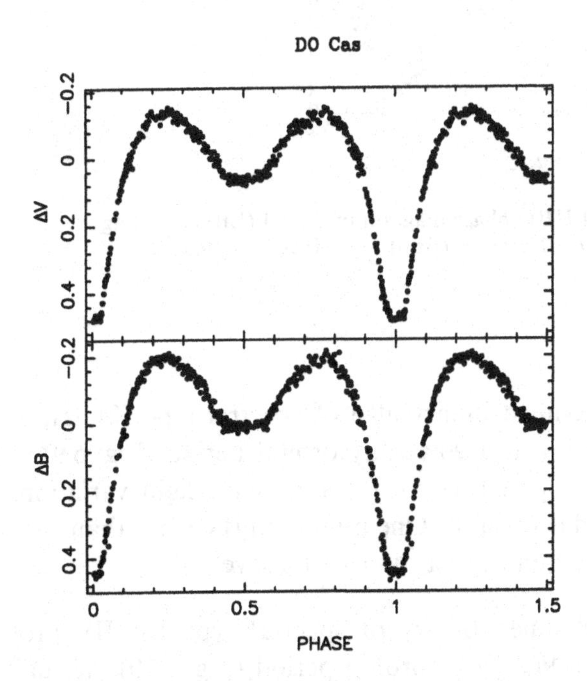

Figure 6.7 Differential V, B phase diagram of DO Cas (EB) according to ephemeris 2439769.2130 + 0.68480E (Gleim & Winkler 1969). Comparison star: BD +59 548.

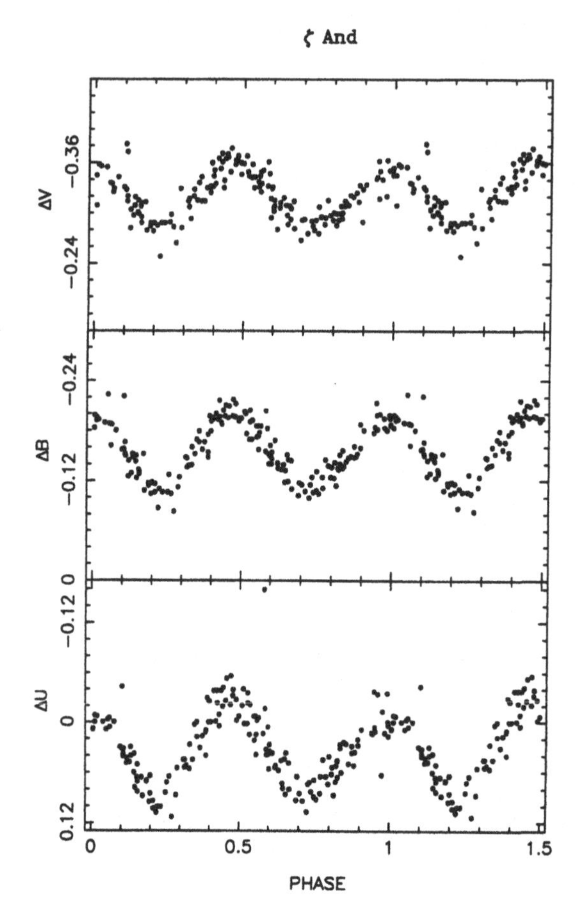

Figure 6.8 Differential *UBV* phase diagram of ζ And (EB) according to ephemeris 2446332.69 + 17.7692*E* (Boyd *et al.* 1990). Comparison star: HR 271.

(iv) AP Piscium: a single-lined binary also of spectral type K1 III, almost surely semi-detached, with a 96.41-day orbital period (Fig. 6.9). There is no evidence that eclipses account for any of the light variation. It is primarily an ellipsoidal variable. One minimum is deeper than the other because of the pointed-end effect discussed above.

(v) UU Cancri: a single-lined binary of spectral type K4 III, probably semi-detached, with a 96.71-day orbital period (Fig. 6.10). Virtually all of the light variation is produced by the ellipticity effect, so it may not be eclipsing at all. Again, the pointed-end effect causes one minimum to be deeper than the other.

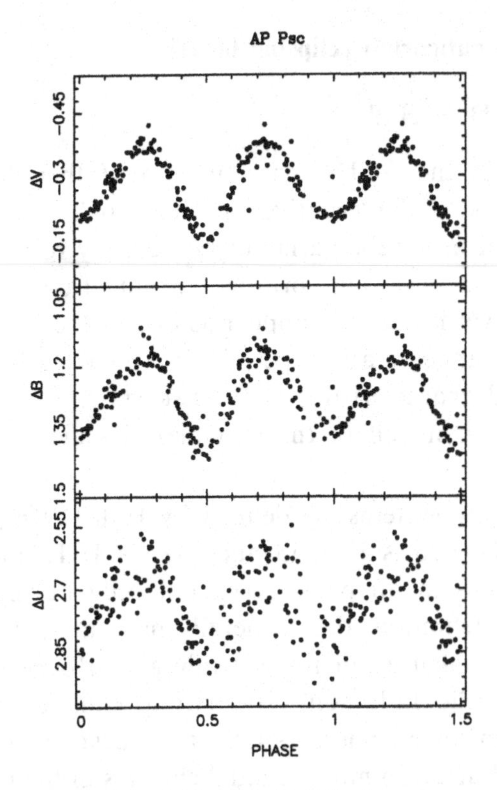

Figure 6.9 Differential *UBV* phase diagram of AP Psc (EB) according to ephemeris 2446249.95 + 96.41*E* (Boyd *et al.* 1990). Comparison star: HD 315.

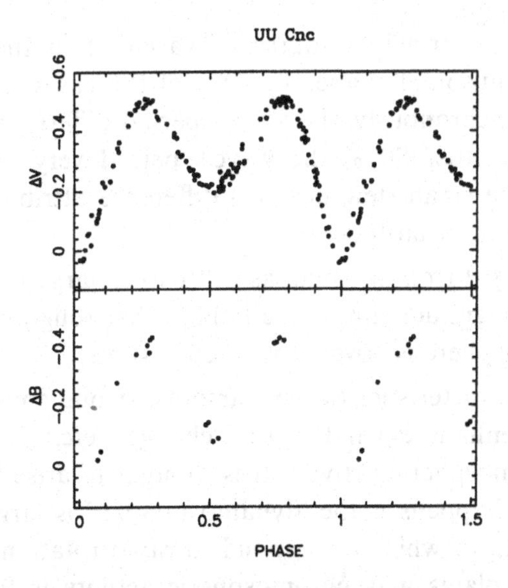

Figure 6.10 Differential *BV* phase diagram of UU Cnc (EB) according to ephemeris 2445031.50 + 96.63*E* (Winiarski & Zola 1987). Comparison star: BD +16 1614.

6.3 RS Canum Venaticorum eclipsing binaries

D. S. Hall

This classification (RS) appears in the *GCVS* twice, in two of their 'main classes'. First, it appears as one type of the 'eruptive variable stars'. This may be misleading because the mechanism for the variability is actually rotational modulation, with the surface brightness non-uniform as a result of cool spots distributed unevenly in longitude, yet it does *not* appear as one of the types of 'rotating variable stars'. Second, it appears as a type of 'close binary eclipsing systems' according to the physical characteristics of the two stars. This also may be misleading because more than half of the variables classified in the *GCVS* as RS are not eclipsing.

The RS Canum Venaticorum type systems, as defined by Hall (1976), are binaries in which the hotter of the two is F or G and the Ca II H and K lines show strong emission reversals at all phases, i.e., not only during an eclipse. Though not part of that definition, they generally have at least one component evolved off the main sequence but not yet filling its Roche lobe, emit intense coronal X-ray and radio radiation, have strong emission lines in the far ultraviolet, lose mass in an enhanced wind, have variable orbital periods, show a starspot wave, and undergo more gradual changes in the mean brightness which may be evidence of a magnetic cycle akin to the sun's 11-year cycle.

The starspot wave, which is the principal source of variation in the RS variables, usually has a nearly sinusoidal shape. In 20% of the cases, where the spotted star is not rotating synchronously, the wave's period differs greatly from the orbital period. In the other 80%, the wave's period very nearly equals the orbital period, with the small (few percent) difference attributable to differential rotation as a function of latitude.

Measured amplitudes have ranged from as small as $0\overset{m}{.}01$ to as large as $0\overset{m}{.}6$ in V (Nolthenius 1991). A half-magnitude amplitude implies that roughly half of the projected area of one hemisphere is covered with dark spots.

The family of RS CVn-type characteristics (large starspots, strong emission lines, intense coronal radiation, enhanced wind, period changes, etc.) is considered the phenomenon of chromospheric activity. It is thought to arise from enhanced dynamo action, which happens if the 'dynamo number' is large or the 'Rossby number' is small, both of which correspond to rapid rotation and a deep convective zone. This explains why chromospheric activity is found in many separately defined groups of both single and binary stars of diverse evolutionary state:

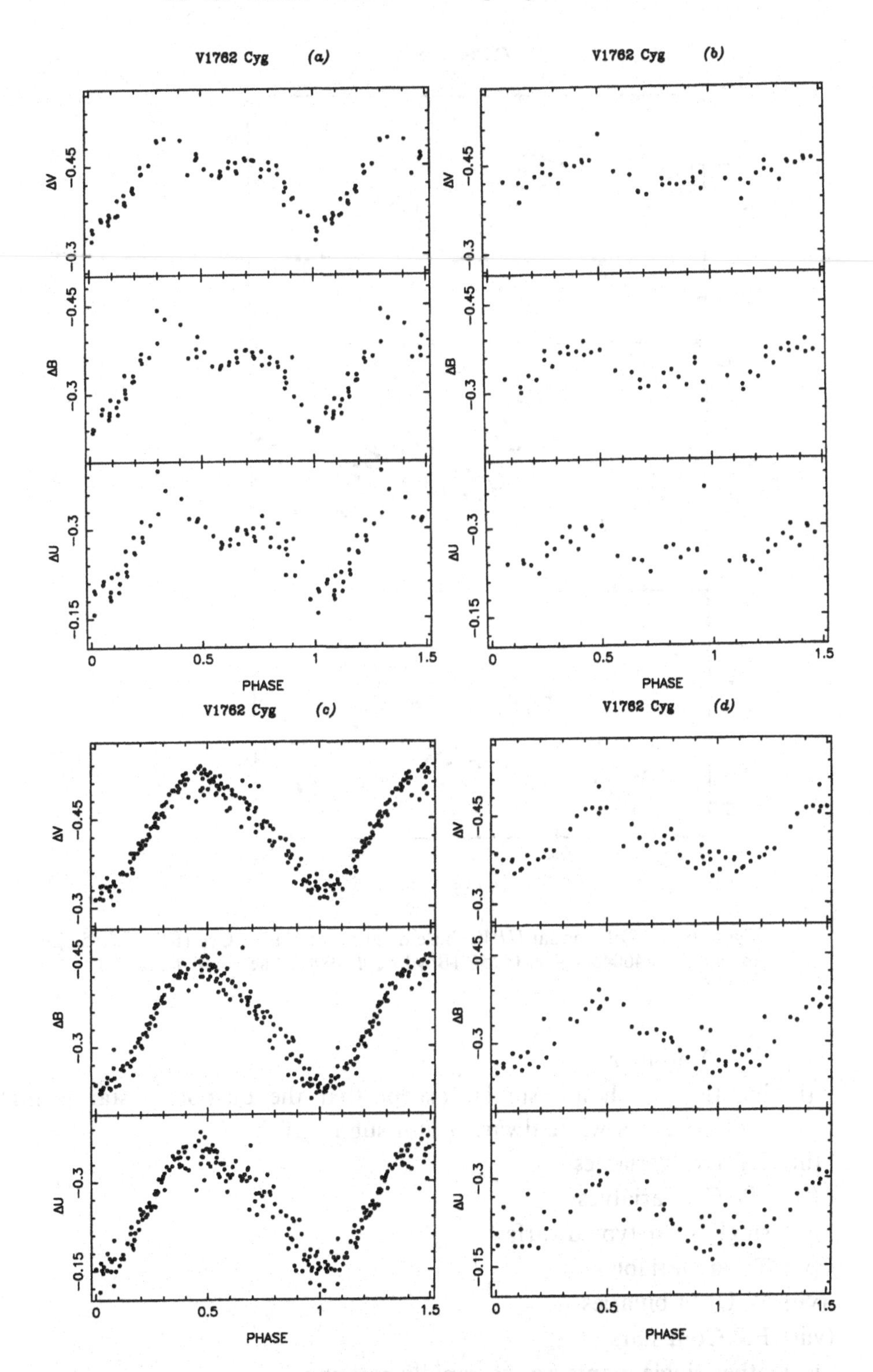

Figure 6.11 Differential *UBV* phase diagram of V 1762 Cyg (RS) according to ephemeris 2445696.0 + 27.875*E* (Boyd *et al.* 1990). Comparison star HD 177483.

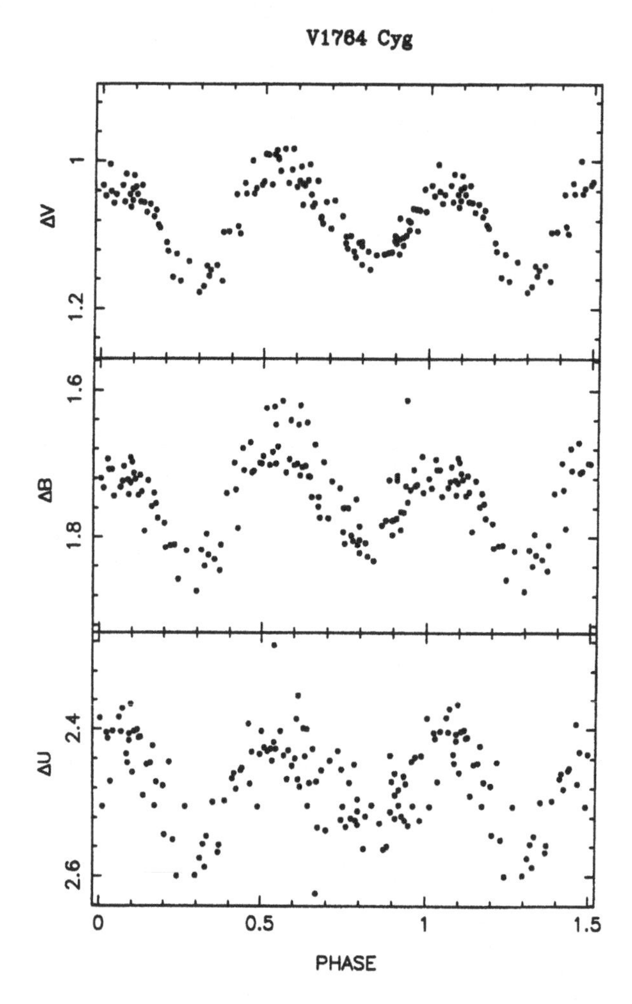

Figure 6.12 Differential *UBV* phase diagram of V 1764 Cyg (RS) according to ephemeris 2446045.8 + 40.1425*E* (Boyd *et al.* 1990). The comparison star is HD 185269.

(i) RS CVn binaries
(ii) binaries which are similar except that the unspotted star is highly evolved, i.e., a white dwarf or hot subdwarf
(iii) BY Dra variables
(iv) UV Cet variables
(v) single solar-type dwarfs
(vi) T Tau variables
(vii) W UMa binaries
(viii) FK Com stars
(ix) other single giants not so rapidly rotating,

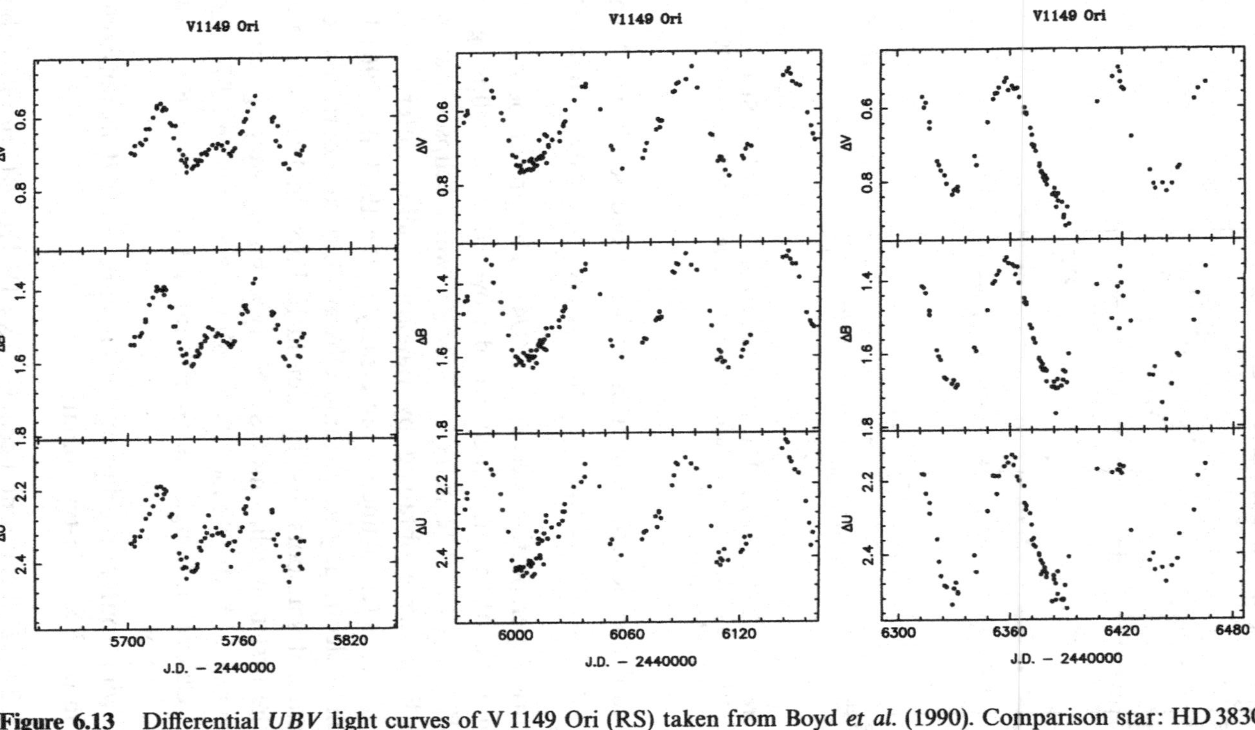

Figure 6.13 Differential *UBV* light curves of V 1149 Ori (RS) taken from Boyd *et al.* (1990). Comparison star: HD 38309.

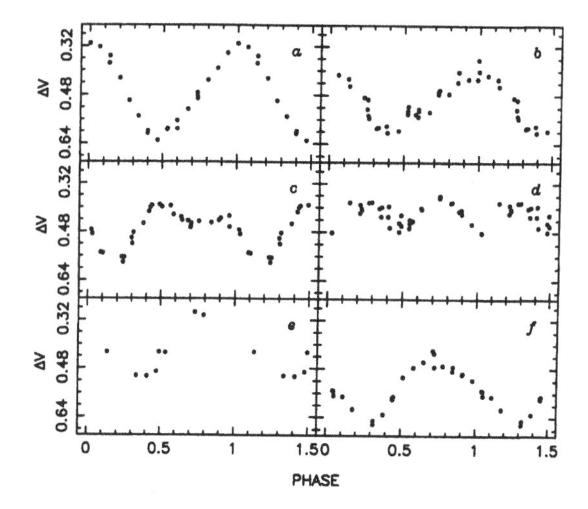

Figure 6.14 Differential B, V phase diagrams of DM UMa (RS). Comparison star: BD +60 1301. Phases are according to ephemeris 2443881.4 + 7.492E (Kimble *et al.* 1981, Mohin *et al.* 1985).

(x) the cool, contact component in semi-detached Algol-type binaries, and
(xi) perhaps the cool component in old novae and cataclysmic variables.

For historical perspective, see Hall (1994). The prototype of the RS type is RS Canum Venaticorum, which was discovered as an eclipsing variable in 1914 by W. Ceraski. The starspot wave between eclipses was seen in the 1920–1922 light curve obtained by Bancroft Walker Sitterly with a visual polarizing photometer, defined more clearly in the 1963 and 1964 light curves obtained photoelectrically at Catania Observatory, and correctly identified as a consequence of starspots first by Gerald E. Kron in 1952 and later by Hall in 1972. The first non-eclipsing RS CVn-type binary varying only as a result of starspots was λ Andromedae. Its variability was discovered in the 1930s by William A. Calder, although the *GCVS* did not classify it RS until 1985, in their fourth edition.

The following four non-eclipsing systems are selected to illustrate starspot variability in the RS CVn-type binaries:

(i) V 1762 Cyg: this is a non-eclipsing K1 IV-III SB1 with an orbital period of $28^{d}.59$ (Fig. 6.11). The light curve shows a nearly sinusoidal starspot wave in some years and a double wave in others, indicative of two dark spot regions on opposite hemispheres of the K giant.
(ii) V 1764 Cyg: this is a non-eclipsing K1 III SB1 with an orbital period of $40^{d}.14$ (Fig. 6.12). Though classified RS, the major source of the variation is the ellipticity effect, with an amplitude of $0^{m}.125$ in V. There

is a starspot wave as well, with an amplitude ranging between $0^{\mathrm{m}}.02$ and $0^{\mathrm{m}}.09$ in various years and a period 0.65% shorter than the orbital period.

(iii) V 1149 Ori: this is a non-eclipsing K1 III SB1 with an orbital period of $53^{\mathrm{d}}.58$ (Fig. 6.13). Since the discovery of variability, the light curve has increased in amplitude from $0^{\mathrm{m}}.05$ to $0^{\mathrm{m}}.40$ mag in V, and has displayed a double as well as a single starspot wave.

(iv) DM UMa: this is a non-eclipsing K1 IV-III SB1 with an orbital period of $7^{\mathrm{d}}.492$ (Fig. 6.14). It has shown both a single and a double starspot wave at various epochs. The maximum amplitude is one of the largest ever seen: $0^{\mathrm{m}}.32$ mag in V.

The interested reader can learn more about chromospheric activity and the RS CVn-type binaries in Baliunas & Vaughan (1985), Hall (1976), Hall (1991), Linsky (1980), Sahade & Wood (1978), Strassmeier *et al.* (1993) and Zeilik *et al.* (1979)

6.4 W UMa type variables

H.W. Duerbeck

Eclipsing binary stars of W Ursae Majoris type are characterized by continuous light changes due to eclipses and due to changing aspects of tidally distorted stars. The minima in the light curves are of almost equal depth, indicating similar surface temperatures of the components, and the periods are short, almost exclusively ranging from about 7 hours up to 1 day.

Since the mass ratio of the two components is always different from unity, one would expect that the stars have different surface temperatures if they follow the standard mass–luminosity–radius relation for hydrogen-burning stars on the main sequence. This would lead to light minima of unequal depth, and thus to a classification EB (eclipsing binaries of β Lyrae-type). The W UMa phenomenon is best explained by the assumption that both stars are in contact, and that the more massive component is transferring luminosity to the less massive one via a common envelope, thus equalizing the surface temperatures. This process is quite difficult to model. Theoretical investigations indicate that the 'thermal contact' might be broken from time to time. Evidence for this scenario comes from some EB systems with masses and periods similar to those found in EW systems.

Despite the fact that EW systems cannot directly be compared with single main sequence stars (because both components violate the mass–luminosity relation by the presence of transfer of luminosity), they are useful for the

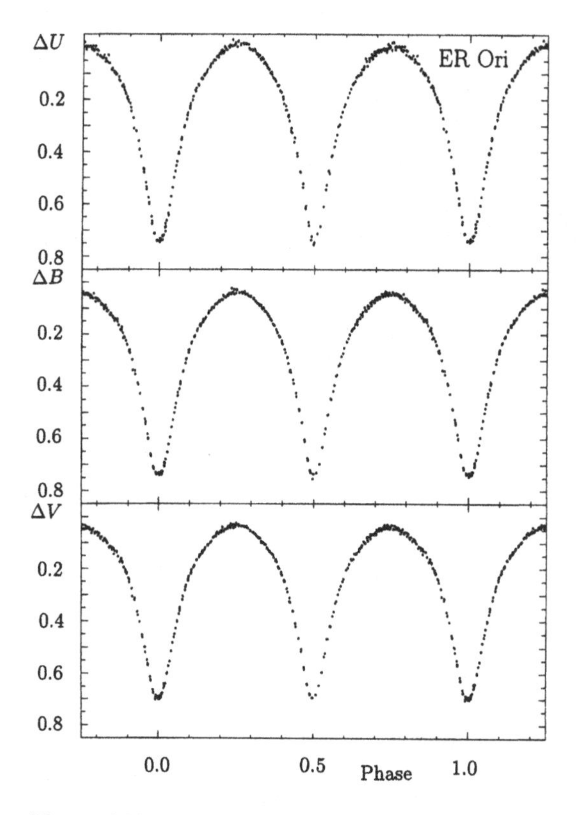

Figure 6.15 Phase diagram of ER Ori based on *UBV* observations carried out at the ESO 50 cm telescope in 1976 by H.W. Duerbeck, according to the ephemeris $2443090.7416 + 0.4233988E$. The magnitudes are differential magnitudes with respect to the comparison star BD -8 1051. A discussion of this system, which contains a physical companion at 0.″19 from the binary, is found in Goecking *et al.* (1994).

derivation of stellar masses, radii and temperatures. The very strong tidal effects lead to ellipsoidal stellar surfaces with varying gravity and emerging luminosity. Light curve synthesis methods (e.g. the programme of Wilson & Devinney 1971) yield reliable results of mass ratio, Roche-lobe fill-out parameter (or fractional radii), inclination and temperature difference. The occurrence of spots on the stellar surfaces can also be taken into account. The derivation of masses and absolute dimensions by spectroscopic observations is hampered by the diffuseness of the spectral lines formed in rapidly rotating contact binaries, therefore cross-correlation methods yield more reliable results than the earlier visual measurements of line positions.

Two subclasses can be distinguished: the A- and the W-type systems. The former are found among the more massive stars of earlier spectral type (A to F) with longer periods, and in these A-type systems the deeper minimum is a

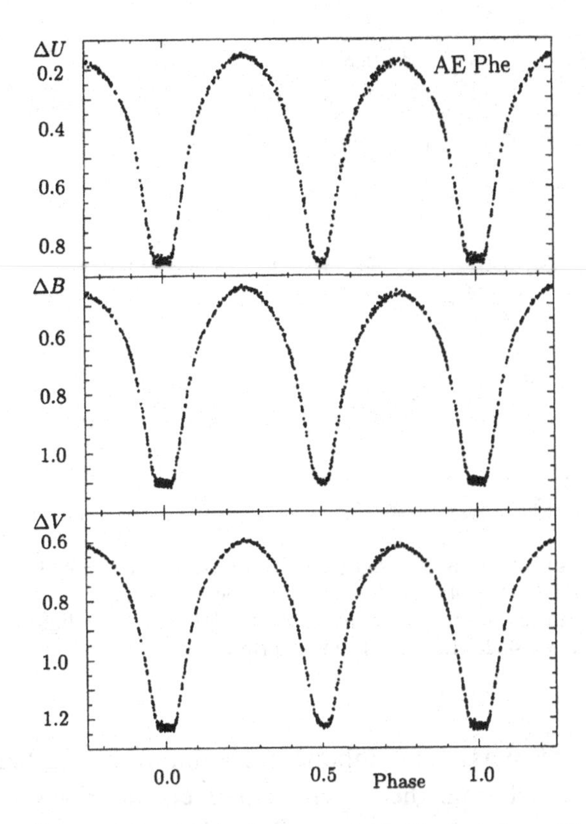

Figure 6.16 Phase diagram of AE Phe based on *UBV* data from Walter & Duerbeck (1988) according to the ephemeris 24426434.8891 + 0.362378*E*. The magnitudes are differential magnitudes with respect to the comparison star HD 9067.

transit – that is, the smaller star has a somewhat lower surface temperature. The latter are found among the less massive systems of later spectral type (G to K) with shorter periods, and in these W-type systems the deeper minimum is an occultation – that is, the smaller star has a somewhat higher surface temperature, a fact for which no convincing explanation has yet been found. Some systems in the intermediate region with unstable light curves can change from an A- to a W-type system in the course of months or years and back again (e.g. TZ Boo, 44 i Boo).

A well-defined period–colour relation exists, with the redder W-type systems having the shorter periods ($0\overset{d}{.}22$–$0\overset{d}{.}4$), while the bluer A-types have longer periods ($0\overset{d}{.}4$–$0\overset{d}{.}8$), see Mochnacki (1985).

Period changes are observed in all EW systems. They are probably connected with the ongoing mass circulation that transports the luminosity from the primary to the secondary. The long-term evolution should lead to a secular

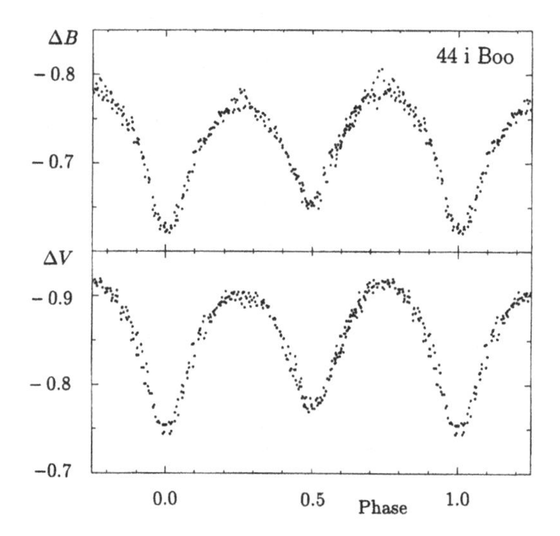

Figure 6.17 Phase diagram of 44 i Boo based on *BV* data from Duerbeck (1978) according to the ephemeris 2442531.4512 + 0.2678159*E*. The magnitudes are differential magnitudes with respect to the comparison star HR 5881. The light curves for JD 2442531.5 and 2442841.5 have been superimposed.

mass loss of the secondary (less massive) component, leading to a lengthening of the period, if no matter is lost from the system. However, the EW systems show a complex behaviour of period changes (period jumps of alternating sign, interrupted by phases of constant period). A recent statistical study by van 't Veer (1991), based on 62 period changes in 29 systems, shows that positive and negative period jumps are randomly distributed, and that simple models of mass transfer from one component to the other (jumps only to one direction) or cyclic magnetic activity (alternating jumps) are not supported.

Due to the low luminosity, EW systems are only known in the immediate solar neighbourhood and – thanks to dedicated search projects – were found also in galactic and globular clusters. EW systems have a high space density. An estimate of the space density of such systems, corrected for selection effects due to the orbital inclination, is 2.5×10^{-5} pc^{-3}, i.e. one out of 500 main sequence stars with spectral types between late A and early K is a W UMa system.

The space distribution of W UMa systems, corrected for selection effects, indicates that they form an old disc population with a typical age of 10^9 years. They likely descend from the short period main sequence RS CVn systems (see Section 6.4) through angular momentum loss by a magnetised stellar wind, and evolve into blue stragglers (old, fairly massive main sequence stars) or into rapidly rotating, spotted giant stars (FK Comae type variables) by merging.

W UMa stars have also been discovered in old stellar aggregates (old galactic clusters and globular clusters).

The first W UMa system discovered was S Ant, whose brightness variation was found by Paul (1888). The *GCVS* lists 542 EW systems. Photoelectric or CCD light curves are available only for a minor fraction, and radial velocity curves exist for an even smaller number of systems.

We selected three systems as examples: ER Ori ($P = 0^{\mathrm{d}}.4234$, spectral type F8), which exhibited a fairly stable light curve at the time of observations (Fig. 6.15), AE Phe ($P = 0^{\mathrm{d}}.3624$, spectral type G0) with a peculiar light curve instability, probably due to spot activity (Fig. 6.16), and 44 i Boo ($P = 0^{\mathrm{d}}.2678$, spectral type G2), a system with a very unstable light curve (Fig. 6.17). For more specific information, we refer to Mochnacki (1985) and Rucinski (1993).

7
X-Ray binaries

J. Krautter

The classification of X-ray binaries is somewhat ambiguous. Some authors consider X-ray binaries to be any kind of interacting close binary with a compact degenerate object – that is, a white dwarf, a neutron star, or a black hole. A more specific definition is that X-ray binaries are only those interacting close binary systems which contain a neutron star or a black hole. In this chapter we shall restrict ourselves to the latter definition; interacting close binaries with a white dwarf are usually called cataclysmic variables which are described in Chapter 5 of this book.

The main (empirical) difference between the cataclysmic variables and the X-ray binaries as defined above is the X-ray luminosity: whereas X-ray binaries have X-ray luminosities of 10^{35}–10^{38} erg s^{-1} (10^{28}–10^{31} W) which corresponds to 25 to 25 000 times the total solar luminosity, cataclysmic variables have $L_X \leq 10^{34}$ erg s^{-1} ($=10^{27}$ W). Hence, X-ray binaries are discovered on the basis of their strong X-ray emission. The basic model of X-ray binaries is a close binary system with a 'normal' star (main sequence or giant, in exceptional cases a degenerate star too) filling its Roche lobe and transferring matter to the compact object, a neutron star or a black hole. Such a system is called a 'semi-detached' system. Due to the orbital angular momentum the matter cannot directly fall onto the compact object, and it forms an accretion disc around the latter (see also Section 5.4). Due to internal friction in the accretion disc (also called viscosity) the matter spirals inward until it eventually falls onto the compact object. In some high-mass X-ray binary systems (see below) where the neutron stars do still have strong magnetic fields, no accretion disc exists; the accretion takes place along the magnetic-field lines onto the magnetic poles of the neutron star (as in the case of the AM Her type cataclysmic variables, Section 5.4). The matter in the parts close to the neutron star is strongly heated up to temperatures of 10 000 000 K or more, and most of the energy is radiated away as thermal radiation (blackbody and bremsstrahlung) in the X-ray range.

The physical mechanism for the high X-ray luminosities is the transformation of gravitational energy into kinetic energy. Accretion is a very efficient tool for that purpose: the accretion energy E_{acc} which can be released in this process scales as $E_{acc} \propto \frac{M_* m}{R_*}$ with M_* and R_* being the mass and the radius of the accreting object, respectively, and m being the mass of the accreted matter. For a given unit mass of one gram, one obtains, for mass and radius of a neutron star, $E_{acc} = 10^{20}$ erg g^{-1} or $E_{acc} \simeq 0.15\,mc^2$. Hence, from the accretion onto a neutron star, 15% of the rest energy mc^2 can be released, which is twenty times as much as one can gain from nuclear fusion of the same amount of pure hydrogen into helium! In view of these numbers it is easily understandable why X-ray binaries do have such high X-ray luminosities. One also sees that E_{acc} for white dwarfs must be much lower, since the radius of a white dwarf is about a thousand times larger than the radius of a neutron star.

The first X-ray binary, Sco X-1, was detected by Giacconi *et al.* (1962) on June 12, 1962 during a sounding rocket flight. For a long time the nature of this object was totally unclear; if the X-ray source was a nearby star, its X-ray luminosity would be 10^7–10^8 times the solar X-ray luminosity, and such objects were not known in the early sixties. A special problem was the large errors in the X-ray positions which for a couple of years prevented an optical identification; eventually Sco X-1 was identified with a faint variable blue star of $V \sim 13$. The most unusual result was the very high ratio of the X-ray to optical luminosity, viz. $L_X/L_{opt} \sim 1000$. In the meantime more than 100 X-ray binaries have become known, and with a few exceptions only, the optical counterparts could be identified too. The idea that X-ray binaries might be close interacting binaries was first put forward in 1967 by the Russian astrophysicist I. Shklovski (Shklovski 1967).

The X-ray binaries can be subdivided in two distinct populations, the low-mass X-ray binaries (LMXRB) and the high-mass X-ray binaries (HMXRB), which are sometimes also called massive X-ray binaries. The subdivision is most clearly seen in the ratio of the X-ray to optical flux. For HMXRBs L_X/L_{opt} is between 10^{-3} and 10; most HMXRBs emit more radiation in the optical spectral range than in the X-ray range; they are bright optical objects. For LMXRBs the situation is the other way around, L_X/L_{opt} is between 10 and 10^4; they are very faint objects in the optical spectral range.

High-mass X-ray binaries

In the HMXRBs the mass-donating companion is an optically bright star of early spectral type; according to its nature, HMXRBs are again subdivided in two groups. In the first group the mass-donating star is an early type (earlier than B3) OB giant or supergiant; in the other one the mass-donating star is a Be star – that is, a B star which exhibits emission lines in its spectrum, see Section 3.3. Accordingly, the members of this latter group – of which slightly more than 20 objects are known by now – are also called Be/X-ray binaries. HMXRBs are very young objects, their spatial distribution shows a strong concentration towards the galactic plane. Accretion probably takes place via a strong stellar wind or via episodic bursts of mass loss in the case of some Be/HMXRBs. One of the main characteristics of the HMXRBs is the presence of periodic X-ray pulses with periods between $0\overset{s}{.}061$ and 835 s which are caused by some kind of lighthouse effect. The accretion of matter takes place along the field lines of the strong neutron-star magnetic field onto the magnetic pole(s). Since the magnetic and the rotation axis of the neutron star are tilted with respect to each other, the magnetic pole is periodically hidden from us; in that case we don't see any X-ray radiation. If the magnetic pole is pointing towards the observer an X-ray pulse can be observed. Hence, the pulse period represents the rotation period of the neutron star.

The optical appearance of the HMXRBs is dominated by the emission from the bright companion whose properties are only slightly influenced by the presence of the X-ray source. Although the variability is most conspicuous in the X-ray spectral range, photometric variability in the optical spectral range is definitely one of the characteristics of the HMXRBs too, even if the amplitude of the variability usually does not exceed 0.1–0.2 mag. Several different physical mechanisms contribute to the observed photometric variability.

The most pronounced photometric variations arise due to the orbital motion of the system. In a first approximation the light curve is sinusoidal and shows two maxima and two minima. The photometric variations are so called 'ellipsoidal variations' which result from a distortion of the bright companion in the gravitational field of the compact object. The centrifugal and tidal effects cause the average optical flux from the companion star to depend upon its orientation with respect to the observer and to the compact object. Depending on the orbital phase, the size of the part of the emitting surface which faces the observer, changes. It is smallest at phases $\Phi = 0.0$ and $\Phi = 0.5$, if the compact object is behind (Φ=0.0) or in front (Φ=0.5) of the companion star. The minima are usually of different depths with the minimum at phase Φ=0.5 being deeper – an effect that is caused by gravity darkening, i.e., the temperature of the

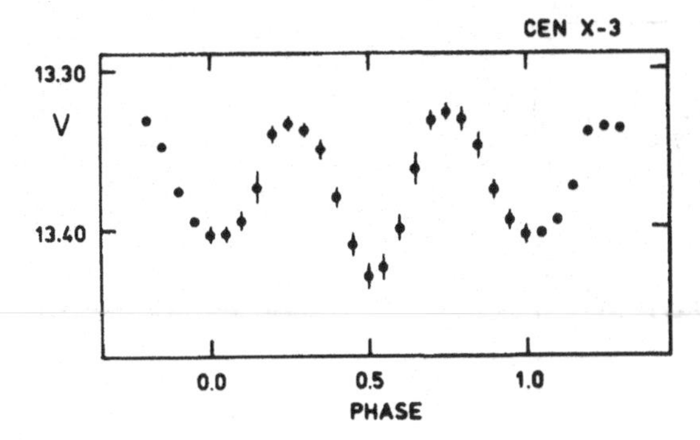

Figure 7.1 Average V phase diagram of Cen X-3 (van Paradijs 1983). Phases are according to ephemeris 244132.081 + 2.08712E.

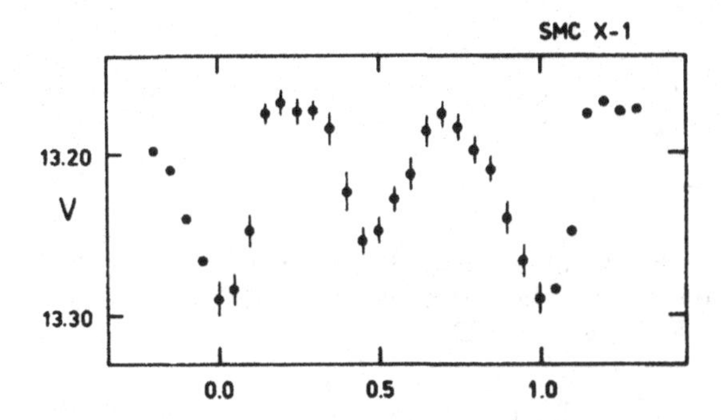

Figure 7.2 Average V phase diagram of SMC X-1 = SK 160 (van Paradijs 1983). Phases are according to ephemeris 2440964.466 + 3.89229E.

surface parts of the companion facing the compact object is lowered due to the strong gravitation.

However, there are several other effects which complicate this picture. First of all, the strong X-ray radiation heats the surface parts of the secondary facing the compact object. In that case one observes a brightening at phase Φ=0.5. In SMC X-1 the X-ray heating is stronger than the gravitational darkening, and hence, the minimum at Φ=0.5 is less deep than the one at Φ=0.0. In most HMXRBs the effect due to the gravitational darkening is stronger than the X-ray heating. In addition, in those cases where we find an accretion disc around the compact object, this accretion disc might contribute on a minor level to the total optical light from the system, showing variations depending on the orbital motion too. The light curves of most HMXRBs show deviations from a pure sinusoidal shape. This is due to the fact that these HMXRBs do not have a circular, but rather a more or less pronounced eccentric orbit.

In addition to the variations caused by the orbital motion, erratic variations on a very minor level below $0^{m}.1$ are found in all HMXRBs. The origin of this

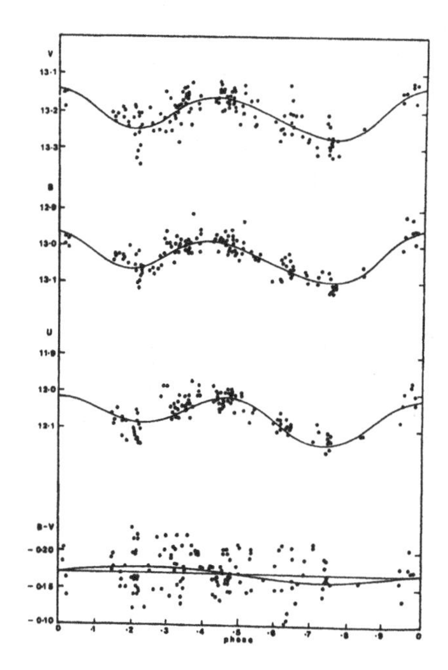

Figure 7.3 *UBV* light and $B - V$ colour curves of SMC X-1 = SK 160. The solid lines are the Fourier fits to the observations (Penfold *et al.* 1975). Phases are according to ephemeris: $2440964.466 + 3.89229E$.

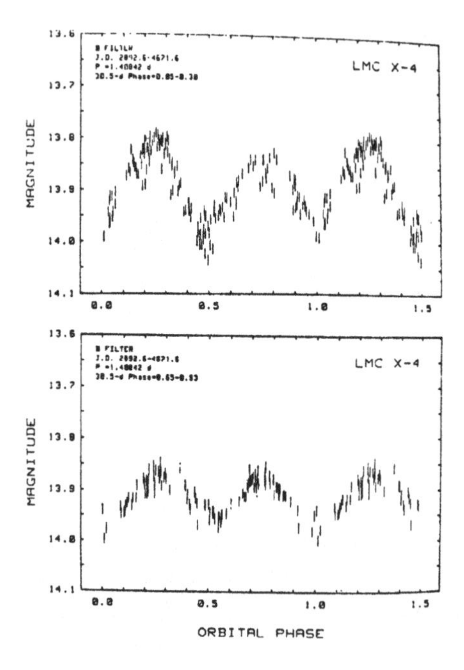

Figure 7.4 Average *B* phase diagram of LMC X-4 with X-rays on (upper part) and X-rays off (lower part) (Chevalier *et al.* 1981). Phases are according to ephemeris $2443215.86 + 1.408E$.

erratic variability is not quite clear; it was suggested that it could be caused by variations in the accretion rate, by semi-regular pulsations, or, as in the case of 4U1700-37/HD 153919, by intrinsic variability in the companion star.

The light curves presented here are a representative sample of HMXRB light curves. In Cen X-3 (Fig. 7.1), which is an average V-light curve of many individual measurements, the minimum at Φ=0.5 is clearly less deep than at Φ=0.0. As already mentioned, in SMC X-1 the X-ray heating is so strong, that it fills in the minimum at Φ=0.5. Again an average light curve is shown, and in addition UBV light curves of SMC X-1 are presented (Figs. 7.2 and 7.3) which clearly show the erratic scatter of the brightness. The last example is LMC X-4 (Fig. 7.4), whose X-ray radiation is periodically switched on and off. It shows two different types of light curves, depending whether the X-rays are on or off. This behaviour can be understood in terms of a tilted accretion disc.

Low-mass X-ray binaries

The structure of the LMXRBs is very similar to the structure of the cataclysmic variables: the companion, a low-mass (in general $M \leq 1M_{\odot}$) late type star, which is filling its Roche-lobe, is transferring matter via an accretion disc onto a neutron star (instead of a white dwarf as in the case of CVs). LMXRBs are intrinsically faint objects in the optical spectral range with an absolute magnitude $V \sim +1$ mag. The spectra of most LMXRBs show a few emission lines superposed on a rather flat continuum. The major fraction of their optical emission comes from the accretion disc and is mainly due to reprocessing of the X-ray radiation in the disc. Reprocessing means that X-ray photons are absorbed in the accretion disc and re-emitted at longer wavelengths. Because of the reprocessed radiation, accretion discs in LMXRBs are generally much brighter than in cataclysmic variables. Accordingly, no brightness variations due to the 'hot spot' have been observed yet, since in the case of LMXRBs the hot spot luminosity is much lower than the total disc luminosity, whereas in cataclysmic variables they are quite often of the same order.

LMXRBs are concentrated towards the galactic center and are also – like the HMXRBs – concentrated towards the galactic plane, but with a wider distribution in latitude. They are also known as galactic X-ray bulge sources. Several of the LMXRBs (> 10) are found in globular clusters; the properties of these objects are indistinguishable from those in the galactic plane. LMXBs are old objects, unlike HMXRBs which are very young. They usually do not show X-ray pulses, which is due to the fact that – because of their high age – their neutron stars no longer have strong magnetic fields; hence, the accretion onto the neutron star takes place from the inner part of the accretion disc and not along magnetic field lines. Many of the LMXRBs produce X-ray bursts during which the X-ray luminosity increases within a few seconds by a factor of 10 or more. These bursts, which re-occur at intervals of hours to days, are caused by thermonuclear events at the surface of the neutron star. If a certain amount of matter has been accreted on the neutron star, thermonuclear burning of helium starts very suddenly, and within a few seconds the X-ray flux increases to the observed values. This process is very similar to the outburst of classical novae; however, in the case of novae the thermonuclear runaway takes place on a white dwarf instead of on a neutron star, and no helium is burned, but hydrogen.

In the case of LMXRBs there is a variety of physical processes which cause photometric variations. As in the case of HMXRBs, periodic variations caused by the orbital motion of the binary components play an important role. By now the orbital periods of some 25–30 LMXRBs are known; they are between 11 min and 9 days. The differences in period length are caused by the nature of the companion star; for periods below 50 min it is a degenerate object, for periods above one day the companion must be an evolved giant star, and for the periods in between, the companion is a main-sequence star. Since the optical variability is caused by varying visibility conditions of the disc, the amplitude of the variability due to orbital motion is correlated with the inclination of the orbital plane of the binary system; it is highest if the orbital plane is in the line of sight, and it is smallest if the orbital plane is perpendicular to the line of sight – that is, if one looks from the poles onto the system. Unlike in HMXRBs, and in some cataclysmic variables, ellipsoidal variations due to the distortion of the companion star cannot be seen in LMXRBs. However, X-ray heating of the companion star seems to play a role in the case of several LMXRBs. Only very few LMXRBs are known to show eclipses. For many years this caused serious doubts about their binary nature, since from statistical reasoning alone one would expect that, as in the case of cataclysmic variables, a significant fraction of the LMXRBs should show eclipses. Now we know that the absence of eclipses is caused by a selection effect: for a system with an inclination $i \geq 75°$ the X-rays are shielded by the accretion disc. On the other hand, since the LMXRBs are found on the basis of their X-ray radiation (one should not forget that they are optically very faint), we simply do not find those systems with high inclination which can only show eclipses. In the meantime a few systems with partial eclipses have been found.

A second very pronounced variability is the burst activity which is primarily observed in the X-ray regime. Simultaneous X-ray and optical observations of bursts have shown that the increase of the optical brightness takes more time than the X-ray rise, the maximum occurs later, and the optical light curves

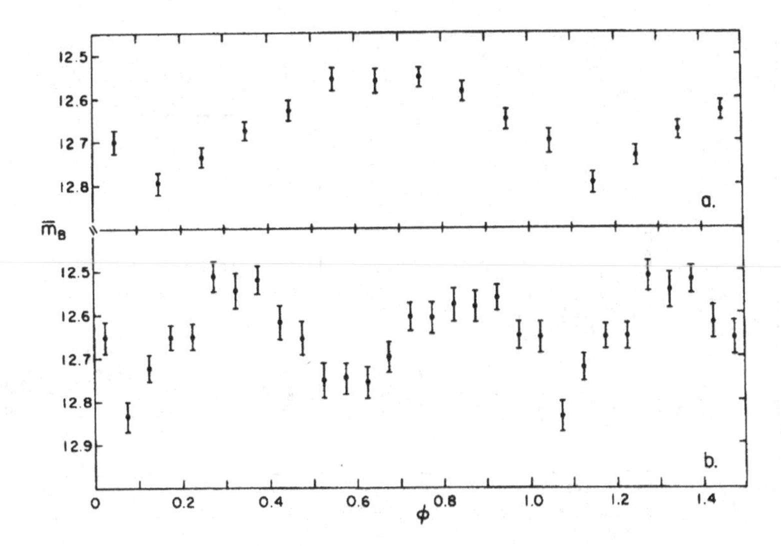

Figure 7.5 *B* phase diagram of V 818 Sco = Sco X-1 based on 1068 photographic and 614 photoelectric magnitude determinations; the error bars show the standard deviations of the mean. Ephemeris for (a) is $2440081.0 + 0.787313E$, for (b) the same epoch and half the value for P was used (Wright *et al.* 1975).

appear to be smeared as compared with the X-ray light curves. This behaviour can be easily understood in terms of the reprocessing model: it takes a few seconds for the X-rays to get from the neutron stars to those places in the discs where the reprocessing occurs. From the differences in the optical and X-ray light curves it is possible to obtain information on the geometric structure of the accretion disc. Typical time scales for the total duration of a burst are of the order of one minute. There is a wide variety of burst profiles, within a given source it can change from one outburst to the next one.

A special group of X-ray binaries are the so called 'transient' sources. They do not show any X-ray activity for a long time, then turn on in a short time (typically a few days), and return to quiescence on a timescale of weeks to months. Because of the rapid increase of the X-ray luminosity they are also called 'X-ray novae'. Also the optical brightness shows a strong increase (5–6 mag) on the same time scale as the X-rays; again reprocessing of the X-rays in the accretion disc is the mechanism for the optical emission. There remain many open questions concerning the explanation of the transient behaviour; it seems to be clear that the sudden onset of the X-ray radiation is due to the onset of mass accretion onto the neutron star. Transient sources in quiescent phases are good candidates to study the companion star, since there is no accretion disc, and hence, the optical light is dominated by radiation from the companion.

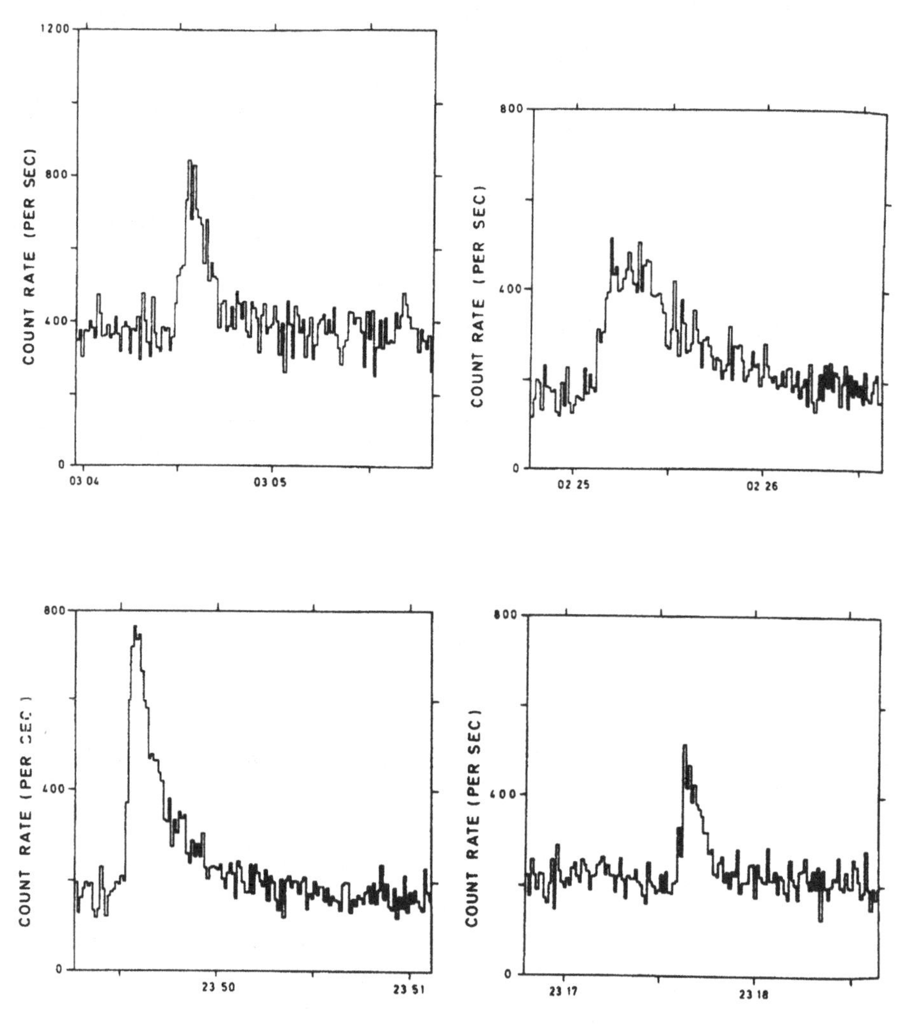

Figure 7.6 Photon count rate per second in white light for V 801 Ara = X 1636-536 (Pedersen *et al.* 1982).

Other photometric variations are low-amplitude periodic pulses which vary with the rotation period of the neutron star. Again the optical pulses are caused by reprocessing of X-ray pulses. In other systems quasi-periodic oscillations occur with timescales between 20 and 1000 s which can amount to up to 50% of the total flux. However, their origin is not very well understood. Also irregular optical variations, the so called flickering, show up; again these are associated with similar X-ray behaviour. For GX 339-4 flickering has been found on timescales as short as 20 msec. In Sco X-1 a study of the brightness since 1900 seems to indicate that there are long-term variations of the average brightness.

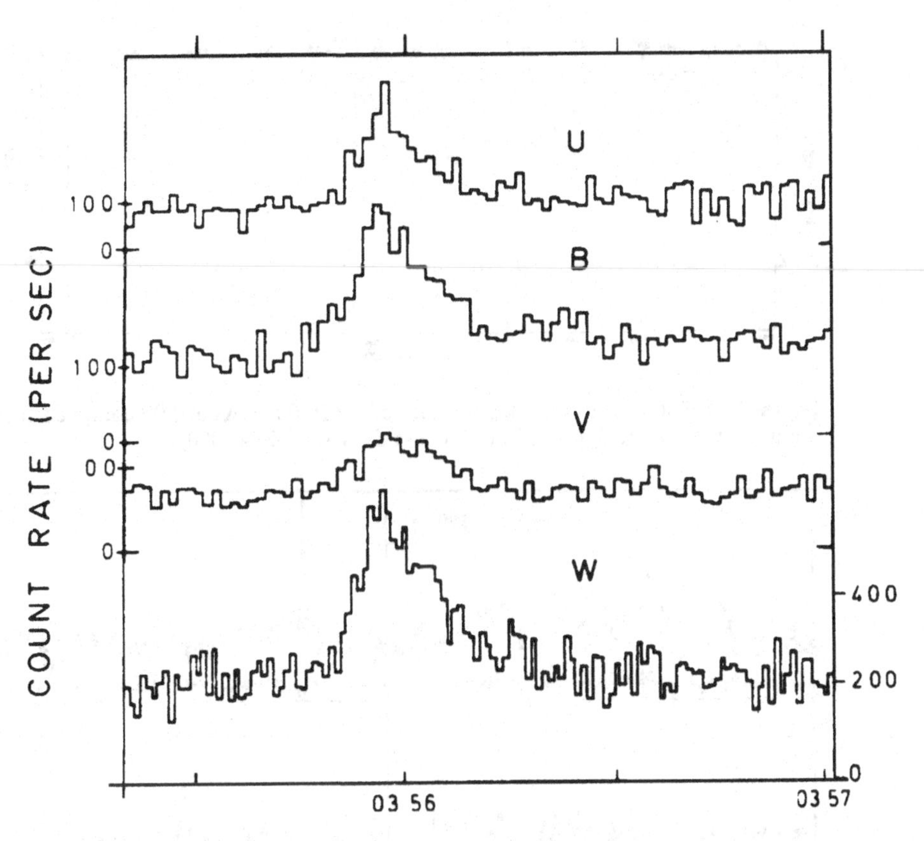

Figure 7.7 Multicolour observations of a single burst of V 801 Ara=X1636-536. The photon count rate per second is given for U, B, V and white light pass bands. Data from Pedersen *et al.* (1982).

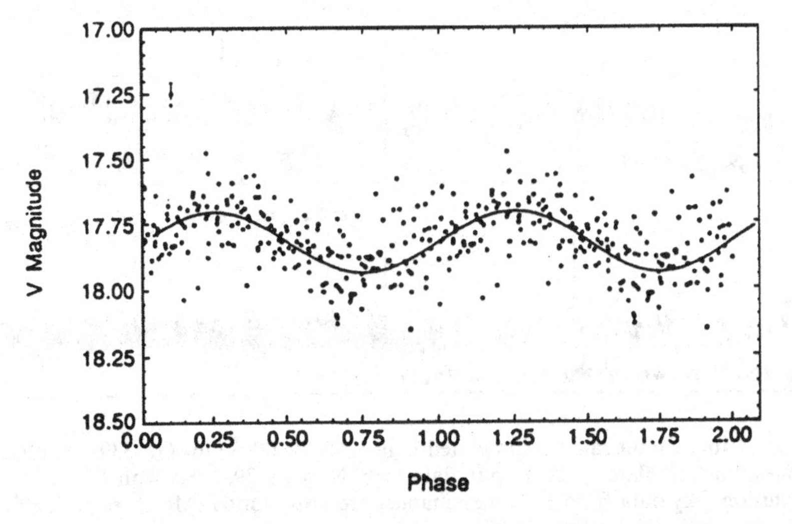

Figure 7.8 Orbital variation of the V magnitude of V 801 Ara. Data from Smale and Mukai (1988).

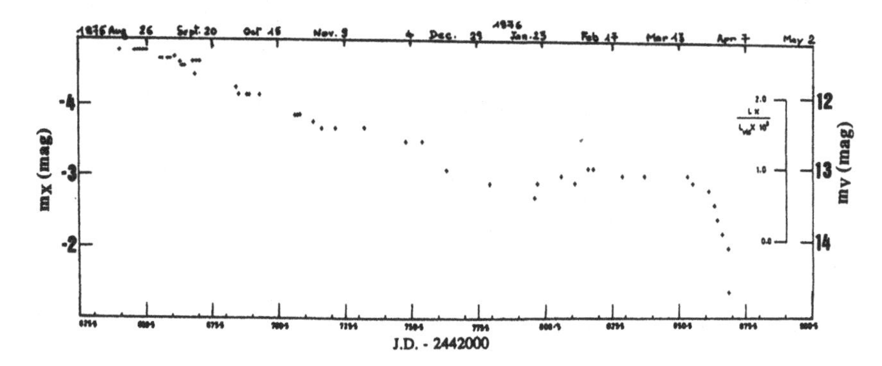

Figure 7.9 Estimated visual magnitudes of X 620-00 = A 0620-00 compiled from the literature by Whelan *et al.* (1977). X-axis is JD − 2 442 000.

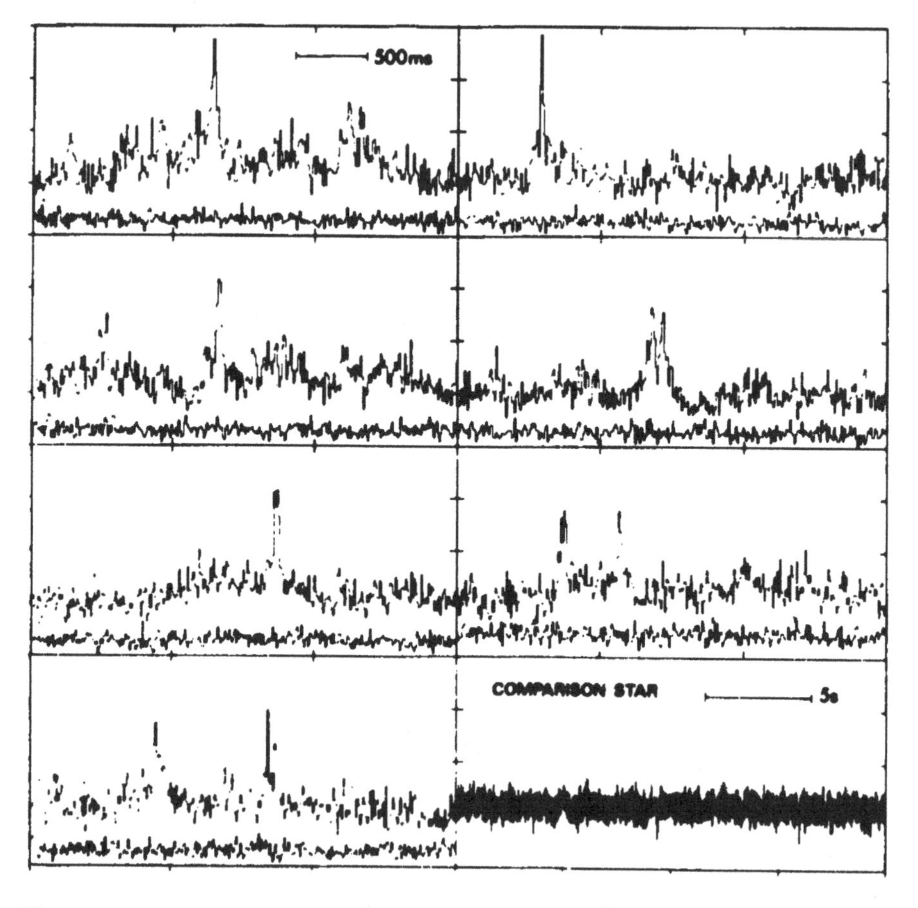

Figure 7.10 Count rates in integrated light (3500–9000 Å) for GX 339-4. Different frames illustrate flares picked up in data from May 28–29, 1980, with 10 ms resolution. Sky data from the other channel are also plotted (Motch *et al.* 1982).

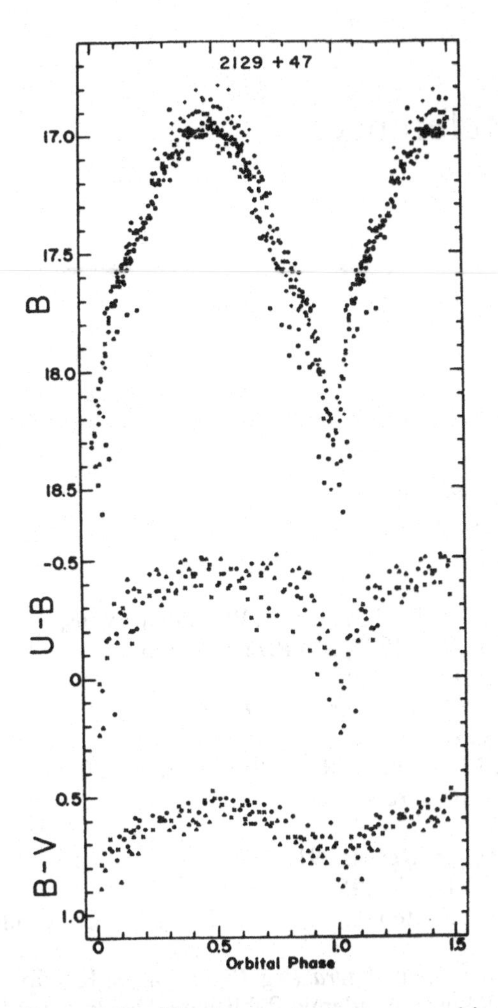

Figure 7.11 Photon count rate per second in *UBV* for V 1727 Cyg = X 2129+47 (McClintock *et al.* 1981).

The LMXRB light curves shown in this chapter display representative examples of the above described photometric behaviour in the optical spectral range. For Sco X-1 both the long-term variations and the orbital variations are shown (Fig. 7.5). For V 801 Ara, the orbital variations, a series of bursts, and multicolour observations of a single burst are shown (Figs. 7.6–7.8). X620-00 is an X-ray nova (Fig. 7.9), GX339-4 (Fig. 7.10) shows very fast optical variations, and V 1727 Cyg is a source which shows partial X-ray eclipses (Fig. 7.11). Its inclination angle is close to 78°, and the optical emission seems to be dominated by X-ray heating of the companion star.

For more general literature, we refer to Bradt & McClintock (1983), Lewin & van den Heuvel (1983) and Mason (1988).

References

Abt, H.A., Cardona, O., 1984, *Ap. J.* 285, 190
Ahnert, P., 1950, *Mitt. Ver. Sterne Sonneberg* 127
Albrecht, S., 1908, *Lick Obs. Bull.* 5, 62
Alcock, C., *et al.*, 1995, *Astron. J.* submitted
Alexander, J.B., Andrews, P.J., Catchpole, R.M., Feast, M.W., Lloyd Evans, T., Menzies, J.W., Wisse, P.N.J., Wisse, M., 1972, *MNRAS* 158, 305
Allen, D. A., 1984, *Proc. Astron. Soc. Aust.* 5, 369
Ammann, M., Hall, D.S., Tate, R.C., 1979, *Acta Astron.* 29, 259
Andersen, J., 1991, *Astr. Ap. Rev.* 3, 91
Andersen, J., Nordström, B., Mayor, M., Polidan, R.S., 1988, *Astr. Ap.* 207, 37
André, Ch., 1899, *Traité d'Astronomie stellaire* 1, 303
Andrews, P.J., 1974, *MNRAS* 167, 635
Appenzeller, I., Mundt, R., 1989, *Astr. Ap. Rev.* 1, 291
Ardeberg A., de Groot M., 1973, *Astr. Ap.* 28, 295
Baade, D., 1987, in *Physics of Be stars*, Slettebak, A., Snow, T.P. (eds.), Cambridge University Press, Cambridge 361
Baade, D., 1992a, in *Evolutionary Processes in Interacting Binary Stars*, Kondo, Y., Sistero, R., Polidan, R.S. (eds.), Kluwer Academic Publishers, Dordrecht, 147
Baade, D., 1992b, in *The Atmospheres of Early-Type Stars*, Huber, U., Jeffery, C.S. (eds.) Springer Verlag, Berlin, 145
Baade, D., Balona, L.A., 1994, in *Pulsation, Rotation and Mass Loss in Early-type Stars*, Balona, L.A., Henrichs, H.F., Le Contel, J.M., (eds.) Kluwer Academic Publishers, Dordrecht, 311
Bahner K., Hiltner, W.A., 1961, *Ap. J. Suppl.* 6, 319
Bahyl, V., Pikler, J., Kreiner, J.M., 1979, *Acta Astron.* 29, 393
Baker, N., Kippenhahn, R., 1962, *Zeitschr. f. Astrophys.* 54, 114
Baliunas, S.L., Vaughan, A.H., 1985, *Ann. Rev. Astr. Ap.* 23, 379
Balog, N.I., Goncharskii, A.V., Cherepaschuk, A.M., 1981, *Soviet Astr.* 25, 38
Balona, L.A., 1990, *MNRAS* 245, 92
Balona, L.A., 1991, in *Rapid Variability of OB-stars: Nature and Diagnostic Value*, ESO Workshop, Baade, D. (ed.) 249
Balona, L.A., Cuypers, J., 1993, *MNRAS* 261, 1
Balona, L.A., Rozowski, J., 1991, *MNRAS* 251, 66P
Balona, L.A., Marang, F., Monderen, P., Reiterman, A., Zickgraf, F.-J., 1987, *Astr. Ap. Suppl.* 71, 11

Balona, L.A., Egan, J., Marang, F., 1989, *MNRAS* 240, 103
Balona, L.A., Cuypers, J., Marang, F., 1992, *Astr. Ap. Suppl.* 92, 533
Barba, R., Niemela, V., 1994, *IAU Circ.* 6099
Barbon, R., Cappellaro, E., Turatto, M., 1989, *Astr. Ap. Suppl.* 81, 421
Barnes, T.G., 1973, *Ap. J. Suppl.* 25, 369
Bateson, F., 1974, *Publ. RASNZ*, VSS 2, 1
Bateson, F., 1977, *New Zealand Journal of Science* 20, 73
Bateson, F., 1994, *IAU Circ.* 6102
Bateson, F., Jones, A., 1957, *Circ. RAS New Zealand Var. Star Sect.* N. 79
Bath, G.T., Clarke, C.J., Mantle, V.J., 1986, *MNRAS* 221, 269
Batten, A.H. (ed.) 1989, *Algols, Ap. Sp. Sci.* 50
Beals, C.S., 1931, *MNRAS* 91, 966
Beals, C.S., 1938, *Trans. IAU* 6, 248
Beaulieu, J.P., *et al.*, 1995, *Astr. Ap.* submitted
Beech, M., 1985, *Ap. Sp. Sci.* 117, 69
Bellingham, J.G., Rossano, G.S., 1980, *Astron. J.* 85, 555
Belyakina, T.S., 1970, *Astrofizika* 6, 49
Bertout, C, 1989, *Ann. Rev. Astr. Ap.* 27, 351
Bibo, E.A., Thé, P.S., 1991, *Astr. Ap. Suppl.* 89, 319
Bode, M., Evans, A. (eds.) 1989, *Classical novae*, J. Wiley, Chichester
Bolton, C.T., 1981, in *Be Stars*, Jaschek, M., Groth, H.G. (eds.), Reidel Academic Publishers, Dordrecht, 181
Bopp, B.W., Fekel, F.C., 1977, *Astron. J.* 82, 490
Bopp, B.W., Moffett, T.J., 1973, *Ap. J.* 185, 239
Bopp, B.W., Stencel, R.E., 1981, *Ap. J. Lett.* 247, L131
Bopp, B.W., Hall, D.S., Henry, G.W., Noah, P., Klimke, A., 1981, *Publ. Astron. Soc. Pac.* 93, 504
Bopp, B.W., Goodrich, B.D., Africano, J.L., Noah, P.V., Meredith, R.J., Palmer, L.H., Quigley, R.J., 1984, *Ap. J.* 285, 202
Bouvier, J., Bertout, C., 1989, *Astr. Ap.* 211, 99
Boyd, L.J., Genet, R.M., Hall, D.S., Busby, M.R., Henry, G.W., 1990, *IAPPP Comm.* 42, 44
Bradt, H.V.D., McClintock, J.E., 1983, *Ann. Rev. Astr. Ap.* 21, 13
Breger, M., 1977, *Publ. Astron. Soc. Pac.* 89, 339
Breger, M., 1988, *Publ. Astron. Soc. Pac.* 100, 751
Breysacher, J., Perrier, C., 1980, *Astr. Ap.* 90, 207
Breysacher, J., Perrier, C., 1991, in *Wolf-Rayet Stars and Interrelations with Other Massive Stars in Galaxies*, van der Hucht, K.A., Hidayat, B. (eds.), Kluwer Academic Publishers, Dordrecht, 229
Burki, G., Heck, A., Bianchi, L., Cassatella, A., 1982, *Astr. Ap.* 107, 205
Buta, R.J., 1982, *Publ. Astron. Soc. Pac.* 94, 578
Buta, R.J., Smith, M.A., 1979, *Ap. J.* 232, 213
Buta, R.J., Turner, A., 1983, *Publ. Astron. Soc. Pac.* 95, 72
Byrne, P.B., Rodonó, M. (eds.) 1983, in *Activity in Red Dwarf Stars*, Reidel Academic Publishers, Dordrecht
Caldwell, J.A.R., Coulson, I.M., 1984, *SAAO Circ.* 8, 1
Caldwell, J.A.R., Coulson, I.M., 1985, *SAAO Circ.* 9, 5
Campbell, L., 1955, *Studies of Long Period Variables*, Cambridge, Mass., AAVSO.
Cannizzo, J.K., 1993, *Ap. J.* 419, 318
Cappacioli, M., Della Valle, M., D'Onofrio, M., Rosino, L., 1989, *Astron. J.* 97, 1622

Cappacioli, M., Della Valle, M., D'Onofrio, M., Rosino, L., 1990, *Ap. J.* 360, 63
Cassatella, A., Viotti, R., 1990, *Physics of Classical Novae*, Springer Verlag, Berlin
Catalano, F.A., Leone, F., 1993, *Astr. Ap. Suppl.* 97, 501
Catalano, F.A., Renson, P., 1984, *Astr. Ap. Suppl.* 55, 371
Catalano, F.A., Kroll, F., Leone, F., 1991, *Astr. Ap.* 248, 179
Celis, S., 1977, *Astr. Ap. Suppl.* 29, 15
Chavarria, C., 1979, *Astr. Ap.* 79, L18
Chentsov, E.L., Luud, L., 1990, *Astrofizica* 31, 415
Cherepaschuk, A.M., 1992, in *Evolutionary Processes in Interacting Binary Stars*, Kondo, Y., Sistero, R., Polidan, R.S. (eds.), Kluwer Academic Publishers, Dordrecht, 123
Cherepaschuk, A.M., Efremov, Yu.N., Kurochkin, N.E., Shakura, N.I., Syunyaev, R.A., 1972, *Inf. Bull. Var. Stars* 720
Cherepaschuk, A.M., Eaton, J.A., Khaliullin, Kh.F., 1984, *Ap. J.* 281, 774
Chevalier, C., Ilovaisky, S.A., Motch, C., Pakull, M., Lub, J., van Paradijs, J., 1981, *Space Sci. Rev.* 30, 405
Christy, R.F., 1966, *Ann. Rev. Astr. Ap.* 4, 353
Chu, Y.H., 1991, in *Wolf-Rayet Stars and Interrelations with Other Massive Stars in Galaxies*, van der Hucht, K.A., Hidayat, B. (eds.), IAU Symp. 143, Kluwer Academic Publishers, Dordrecht, 349
Clausen, J.V., 1996, *Astr. Ap.* 308, 151.
Clausen, J.V., Giménez, A., van Houten, C.J., 1995, *Astr. Ap. Suppl.* 109, 425
Cocke, W.J., Disney, M.J., Taylor, D.J., 1969, *Nature* 221, 525
Cohen, J.G., 1985, *Ap. J.* 292, 900
Connelley, M., Sandage, A.R., 1967, *Publ. Astron. Soc. Pac.* 70, 600
Conti, P.S., 1984, in *Observational Tests of Stellar Evolution Theory*, Maeder, A. (ed.), Reidel Academic Publishers, Dordrecht, 233
Cousins, A.W.J., 1951, *MNASSA* 10, 60
Cousins, A.W.J. , 1958, *MNASSA* 17, 132
Cox, A.N., 1993, in *New Perspectives on Stellar Pulsation and Pulsating Variable Stars*, IAU Coll. 139, Nemec, J.M., Matthews, J.M. (eds.), Cambridge University Press, Cambridge, 107
Cox, J.P., 1980, *The Theory of Stellar Pulsation*, Princeton University Press
Cramer, N., 1994, in *Applications de la photométrie de Geneève aux étoiles B et à l'extinction interstellaire*, Thèse 2692, Université de Genève
Crews, L.J., *et al.*, 1995, *Astron. J.* in press
Cristaldo, S., Rodonó, M., 1973, *Astr. Ap. Suppl.* 10, 47
Cropper, M., 1985, *MNRAS* 212, 709
Cuypers, J., 1991, in *Rapid Variability of OB-stars: Nature and Diagnostic Value*, ESO Workshop, Baade, D. (ed.) 83
Cuypers, J., Balona, L.A., Marang, F., 1989, *Astr. Ap. Suppl.* 81, 151
Davidson, K., 1989, in *Physics of Luminous Blue Variables*, Davidson, K., Moffat, A.F.J., Lamers, H.J.G.L.M. (eds.), Kluwer Academic Publishers, Dordrecht, 101
de Groot, M., 1969, *Bull. Astron. Inst. Neth.* 20, 225
de Groot, M., 1988, *Ir. Astron. J.* 18, 163
de Jager, C., 1980, *The Brightest Stars*, Reidel Academic Publishers, Dordrecht
de Jager, C., van Genderen, A.M., 1989, in *Physics of Luminous Blue Variables*, Davidson, K., Moffat, A.F.J., Lamers, H.J.G.L.M. (eds.), Kluwer Academic Publishers, Dordrecht, 127

de Koter, A., 1993, in *Studies of the Variability of Luminous Blue Variables*, Ph.D. Thesis, Utrecht, The Netherlands

de La Caille, N.L., 1763, *Caelum Australum Stelliferum*, Paris

de Vaucouleurs, G., Buta, R., Ables, H.D., Hewitt, A.V., 1981, *Publ. Astron. Soc. Pac.* 93, 36

Dean, C.A., 1974, *Publ. Astron. Soc. Pac.* 86, 912

Dean, J.F., Cousins, A.W.J., Bywater, R.A., Warren, P.R., 1977, *Mem. R. Astron. Soc.* 83, 69

Dickel, H.R., Lortet M.-C., de Boer, K.S., 1987, *Astr. Ap. Suppl.* 68, 75

Diethelm, R., 1983, *Astr. Ap.* 124, 108

Diethelm, R., 1990, *Astr. Ap.* 239, 186

Downes, R.A., Shara, M.M., 1993, *Publ. Astron. Soc. Pac.* 105, 127

Duerbeck, H.W., 1978, *Astr. Ap. Suppl.* 32, 361

Duerbeck, H.W., 1981, *Publ. Astron. Soc. Pac.* 83, 165

Duerbeck, H.W., 1987, *A Reference Catalogue and Atlas of Galactic Novae*, Reidel Academic Publishers, Dordrecht (= *Space Sci. Rev.* 45, 1)

Duerbeck, H.W., Seitter, W.C., 1982, in *Landolt-Börnstein Numerical Data and Functional Relationships in Science and Technology*, New Series Group VI, Vol. 2b, Schaifers, K., Voigt, H.H. (eds.), Springer, Berlin, 197

Duerbeck, H.W., Seitter, W.C., 1995, in *Landolt-Börnstein Numerical Data and Functional Relationships in Science and Technology*, in press

Dziembowski, W.A., 1994, in *Pulsation, Rotation and Mass Loss in Early-type Stars*, Balona, L.A., Henrichs, H.F., Le Contel, J.M., (eds.), Kluwer Academic Publishers, Dordrecht, 55

Dziembowski, W.A., Pamyatnykh, A.A., 1993, *MNRAS* 262, 204

Eggen, O.J., 1970, *Ap. J. Suppl.* 22, 289

Eggen, O.J., 1975, *Ap. J. Suppl.* 29, 77

Eggen, O.J., 1977, *Ap. J. Suppl.* 34, 233

Engelbrecht, C.A., Balona, L.A., 1986, *MNRAS* 219, 449

Evans, D.S., 1959, *MNRAS* 119, 526

Faulkner D.J., Shobbrook, R.R., 1979, *Ap. J.* 232, 197

Feast, M.W., 1975, in *Variable Stars and Stellar Evolution*, Sherwood, V.E., Plaut, L. (eds.), Reidel Academic Publishers, Dordrecht, 129

Feast, M.W., 1995, in *Astrophysical Applications of Stellar Pulsation*, Stobie, R.S., Whitelock, P.A., (eds.), *Astr. Soc. Pacif.* in press

Feast, M.W., Thackeray, A.D., Wesselink, A.J., 1960, *MNRAS* 121, 337

Fekel, F.C., Moffett, T.J., Henry, G.W., 1986, *Ap. J. Suppl.* 60, 551

Fernandez, A., Lortet, M.-C., Spite, F., 1983, *Astr. Ap. Suppl.* 52, 4

Fernie, J.D., 1985, *Publ. Astron. Soc. Pac.* 97, 653

Florentin Nielsen, R., 1983, *Report Inst. Theor. Astrophys. Oslo*, 59, 141

Freedman, W.L., *et al.*, 1994, *Nature* 371, 757

Frost, E.B., 1902, *Ap. J.* 15, 340

Frost, E.B., 1906, *Ap. J.* 24, 259

Gahm, G.F., Fischerström, C., Liseau, R., Lindroos, K.P., 1989, *Astr. Ap.* 211, 115

Gallagher, J.S., Starrfield, S., 1978, *Ann. Rev. Astr. Ap.* 16, 171

Gaposchkin, S., 1952, *Harvard Annals* 118, 171

Gautschy, A., Saio, H., 1993, *MNRAS* 262, 213

Gehrz, R.D., 1988, *Ann. Rev. Astr. Ap.* 26, 377

Giacconi, R., Gursky, H., Paolini, F.R., Rossi, B., 1962, *Phys. Rev. Lett.* 9, 439

Gieren, W., 1981, *Ap. J. Suppl.* 47, 315

Gies, D.R., 1991, in *Rapid Variability of OB-stars: Nature and Diagnostic Value*, ESO Workshop, Baade, D. (ed.) 229
Gingold, R., 1984, *Mem. Soc. Astron. Ital.* 56, 169
Gleim, J.K., Winkler, L., 1969, *Astron. J.* 74, 1191
Goecking, K.-D., Duerbeck, H.W., Plewa, T., Kaluzny, F., Schertl, D., Weigelt, G., Flin, P., 1994, *Astr. Ap.* 289, 827
Gosset, E., Rauw, G., Vreux, J.M., Manfroid, J., Sterken, C., 1994, in *The Impact of Long-Term Monitoring on Variable-Star Research*, NATO ARW, Sterken, C., de Groot, M. (eds), NATO ASI Series C, 436, Kluwer Academic Publishers, Dordrecht, 101
Gould, B.A., 1879, *Res. Obs. Nac. Argen.* 1, 1
Gratton, L., 1963, in *Stellar Evolution*, Gratton, L. (ed.), Academic Press, New York, 297
Gray, D.F., 1989, *Ap. J.* 347, 1021
Grison P., *et al.*, 1994, *Astr. Ap.* 289, 404
Guthnick, P., 1913, *Astron. Nachr.* 196, 357
Hack, M., La Dous, C., 1993, *Cataclysmic Variables and Related Objects*, NASA SP-507, NASA/CNRS Monograph Series, Paris/Washington
Haisch, B., Strong, K.T., Rodonó, M., 1991, *Ann. Rev. Astr. Ap.* 29, 275
Halbwachs, 1988, Xième Journée de Strasbourg, Observatoire de Strasbourg (ed.), 53
Hall, D.S., 1976, in *Multiple Periodic Variable Stars*, Fitch, W.S. (ed.), Reidel Academic Publishers, Dordrecht, 387
Hall, D.S., 1986, *Ap. J. Lett.* 309, L83
Hall, D.S., 1990a, in *Active Close Binaries*, Ibanoglu, C. (ed.), Kluwer Academic Publishers, Dordrecht, 95
Hall, D.S., 1990b, *Astron. J.* 100, 554
Hall, D.S., 1991, *IAU Coll.* 130, 353
Hall, D.S., 1994, *IAPPP Comm.* 54, 1
Hall, D.S., 1995, in *Robotic Telescopes: Current Capabilities, Present Developments, and Future Prospects for Automated Astronomy*, Henry, G.W., Eaton, J.A. (eds.), Astron. Soc. of the Pacific
Hall, D.S., Hardie, R.H., 1969, *Publ. Astron. Soc. Pac.* 81, 754
Hall, D.S., Gertken, R.H., Burke, E.W., 1970, *Publ. Astron. Soc. Pac.* 82, 1077
Hall, D.S., Cannon, R.O., Rhombs, C.R., 1984, *Astron. J.* 87, 559 (IAU archive file 141)
Hamuy, M., Suntzeff, N.B., Gonzalez, R., Martin, G., 1988, *Astron. J.* 95, 63
Hansen, C.J., 1980, in *Nonradial and Nonlinear Stellar Pulsation*, Hill, H.A., Dziembowski, W.A. (eds.), *Lecture Notes in Physics* 125, Springer, Berlin, 445
Harmanec, P., 1983, *Hvar Obs. Bull.* 7, 55
Harmanec, P., 1987, in *Physics of Be stars*, Slettebak, A., Snow, T.P. (eds.), Cambridge University Press, Cambridge, 339
Harmanec, P., 1989, *Bull. Astron. Inst. Czech.* 40, 201
Harmanec, P., 1994, in *The Impact of Long-Term Monitoring on Variable-Star Research*, NATO ARW, Sterken, C., de Groot, M. (eds), NATO ASI Series C, 436, Kluwer Academic Publishers, Dordrecht, 55
Harris, H.C., 1985, in *Cepheids: Theory and Observation*, Madore, B. (ed.), Cambridge University Press, Cambridge 232
Haug, U., 1979, *Astr. Ap.* 80, 119
Heck, A., Jasniewicz, G., Jaschek, C., 1988, in *Astronomy from Large Databases*, ESO Conf. and Workshop Proc. No. 28, Murtagh, F., Heck, A. (eds.), 111

Helfand, D.J., Huang, J.H. (eds.), 1985, in *Origin and Evolution of Neutron Stars*, IAU Symposium 125
Henize, K.G., 1956, *Ap. J. Suppl.* 2, 315
Henize, K.G., 1976, *Ap. J. Suppl.* 30, 491
Henrichs, H.F., Bauer, F., Hill, G.M., Kaper, L., Nichols-Bohlin, J.S., Veen, P.M., 1993, in *New Perspectives on Stellar Pulsation and Pulsating Variable Stars*, Nemec, J.M., Matthews, J.M. (eds.), Cambridge University Press, Cambridge, 186
Hensberge, H., 1993, in *Peculiar Versus Normal Phenomena in A-type and Related Stars*, Proc IAU Coll 138, Dworetsky, M., Castelli, F. (eds.) Kluwer Academic Publishers, Dordrecht,
Hensberge, H., 1994, in *The Impact of Long-Term Monitoring on Variable-Star Research*, NATO ARW, Sterken, C., de Groot, M. (eds), NATO ASI Series C, 436, Kluwer Academic Publishers, Dordrecht, 197
Hensberge, H., Manfroid, J., Sterken, C., 1992, *Messenger* 70, 35
Herbig, G.H., 1977, *Ap. J.* 217, 693
Herbst, W., Holtzman, J.A., Phelps, B.E., 1982, *Astron. J.* 87, 1710
Hertzog, K.P., 1992, *Observatory* 112, 105
Hewish, A., Bell, S.J., Pilkington, J.D.H., Scott, P.F., Collins, R.A., 1968, *Nature* 217, 709
Hillier, D.J., 1992, in *The Atmospheres of Early-type Stars*, Heber, U., Jeffery, C.S. (eds.), Springer Verlag, Berlin, 105
Hoffmeister, C., 1984, *Veränderliche Sterne*, (2nd ed.) J.A. Barth Verlag, Leipzig
Houzeau, J.-C., 1878, *Ann. de l'Observatoire Royal de Bruxelles* I, 1
Howarth, I. D., 1977, *JBAA* 87, 395
Hubble, E.P., Sandage, A., 1953, *Ap. J.* 118, 353
Humphreys, R.M., 1989, in *Physics of Luminous Blue Variables*, Davidson, K., Moffat, A.F.J., Lamers, H.J.G.L.M. (eds.), Kluwer Academic Publishers, Dordrecht 3
Humphreys, R.M., Davidson, K., 1987, *Ap. J.* 232, 409
Humphreys, R.M., Davidson, K., 1994, *Publ. Astron. Soc. Pac.* 106, 1025
Hunger, K., Schönberner, D., Rao, N.K. (eds.), 1986, *Hydrogen Deficient Stars and Related Objects*, Reidel Academic Publishers, Dordrecht
Huth, H., Wenzel, W., 1981, *Bull. Inf. CDS* 20, 205
Huth, H., Wenzel, W., 1992, *Bibliographic Catalogue of Variable Stars* (BCVS), Part I (1992 edition), CDS-Archive VI/67
Ichikawa, S., Hirose, M., Osaki, Y., 1993, *Publ. Astron. Soc. Japan* 45, 243
Iglesias, C.A., Rogers, F.J., 1991, *Ap. J. Lett.* 371, L73
Iglesias, C.A., Rogers, F.J., Wilson, B.G., 1987, *Ap. J. Lett.* 322, L45
Iglesias, C.A., Rogers, F.J., Wilson, B.G., 1990, *Ap. J.* 360, 221
Iglesias, C.A., Rogers, F.J., Wilson, B.G., 1992, *Ap. J.* 397, 717
Innes, R.T.A., 1903, *Cape Ann.* 9, 75B
Jakate, S.M., 1979, *Astron. J.* 84, 1042
Jaschek, C., 1982, ESA SP-177, 133
Jaschek, C. 1989, *Data in Astronomy*, Cambridge University Press, Cambridge
Jaschek, C., Jaschek, M., 1987, *The Classification of Stars*, Cambridge University Press, Cambridge
Jaschek, C., Breger, M., 1988, *Bull. Inf. CDS* 35, 93
Jaschek, M., Slettebak, A., Jaschek, C., 1981, *Be Newsletter* 4, 9
Jasniewicz, G., Duquennoy, A., Acker, A., 1987, *Astr. Ap.* 180, 145
Jasniewicz, G., Acker, A., Mauron, N., Duquennoy, A., Cuypers, J., 1994a, *Astr. Ap.* 286, 211

Jasniewicz, G., Lapierre, G., Monier, R., 1994, *Astr. Ap.* 287, 591
Jerzykiewicz, M., Sterken, C., 1979, in *Changing Trends in Variable Star Research*, (IAU Coll 46), Bateson F.M., Smak J., Urch I.H. (eds.), University of Waikato, 474
Jerzykiewicz, M., Sterken, C., 1984, *MNRAS* 211, 297
Jerzykiewicz, M., Wenzel, W., 1977, *Acta Astron.* 27, 35
Johnson, H.R., Querci, F.R., 1986, in *The M-Type Stars*, NASA/CNRS Monograph Series, Paris/Washington
Jones, R.V., Carney, B.W., Latham, D.W., Kurucz, R.L., 1987, *Ap. J.* 314, 605
Joy, A.H., 1942, *Publ. Astron. Soc. Pac.* 54, 15
Kaper, L., 1994, in *Wolf-Rayet Stars: Binaries, Colliding Winds, Evolution*, IAU Symp. 163, van der Hucht, K.A., Williams, P.M. (eds.), Kluwer Academic Publishers, Dordrecht, in press
Kaye, A.B., *et al.*, 1995, *Astron. J.* in press
Kazarovets, E.V., Samus', N.N., Goranskij, V.P., 1993, *Inf. Bull. Var. Stars* 3840
Kenyon, S.J., 1986, in *The Symbiotic Stars*, Cambridge University Press, Cambridge
Kholopov, P.N., Samus', N.N., Frolov, M.S., Goranskij, V.P., Gorynya, N.A., Kireeva, N.N., Kukarkina, N.P., Kurochkin, N.E., Medvedeva, G.I., Perova, N.B., Shugarov, S.Yu, 1985a, *General Catalogue of Variable Stars*, Nauka Publ. House, Moscow 4th ed., vol. 1
Kholopov, P.N., Samus', N.N., Frolov, M.S., Goranskij, V.P., Gorynya, N.A., Kazarovets, E.V., Kireeva, N.N., Kukarkina, N.P., Kurochkin, N.E., Medvedeva, G.I., Perova, N.B., Rastorguev, A.S., Shugarov, S.Yu., 1985b, *General Catalogue of Variable Stars*, Nauka Publ. House, Moscow 4th ed., vol. 2
Kholopov, P.N., Samus', N.N., Frolov, M.S., Goranskij, V.P., Gorynya, N.A., Karitskaya, E.A., Kazarovets, E.V., Kireeva, N.N., Kukarkina, N.P., Medvedeva, G.I., Pastukhova, E.N., Perova, N.B., Shugarov, S.Yu., 1987a, *General Catalogue of Variable Stars*, Nauka Publ. House, Moscow 4th ed., vol. 3
Kholopov, P.N., Samus', N.N., Kazarovets, E.V., Kireeva, N.N., 1987b, *Inf. Bull. Var. Stars* 3058
Kholopov, P.N., Samus', N.N., Kazarovets, E.V., Frolov, M.S., Kireeva, N.N., 1989, *Inf. Bull. Var. Stars* 3323
Kholopov, P.N., Samus', N.N., Durlevich, O.V., Kazarovets, E.V., Kireeva, N.N., Tsvetkova, T.M., 1990, *General Catalogue of Variable Stars*, Nauka Publ. House, Moscow 4th ed., vol. 4
Kim, T.H., 1980, *Ap. Sp. Sci.* 68, 358
Kimble, R.A., Kahn, S.M., Bowyer, S., 1981, *Ap. J.* 251, 585 (IAU archive file 85)
Klimek, Z., Kreiner, J.M., 1973, *Acta Astron.* 23, 331
Klimek, Z., Kreiner, J.M., 1975, *Acta Astron.* 25, 29
Kukarkin, B.V., Parenago, P.P., Jefremow, J.I., Cholopow, P.N., 1955, *General Catalogue of Variable Stars*, Nauka Publ. House, Moscow 1st ed., 7th Suppl.
Kukarkin, B.V., *et al.*, 1971, *First supplement to the third edition of the General Catalogue of Variable Stars*
Kukarkin, B.V., Kholopov, P.N., Artiukhina, N.M., Federovich, V.P., Frolov, M.S., Goranskij, V.P., Gorynya, N.A., Karitskaya, E.A., Kireeva, N.N., Kukarkina, N.P., Kurochkin, N.E., Medvedeva, G.I., Perova, N.B., Ponomareva, G.A., Samus', N.N., Shugarov, S.Yu., 1982, *New Catalogue of Suspected Variable Stars*, Nauka Publ. House, Moscow
Kurtz, D.W., 1982, *MNRAS* 200, 503
Kurtz, D.W., 1990, *Ann. Rev. Astr. Ap.* 28, 607

Kurtz, D.W., 1991, *MNRAS* 249, 468
Kurtz, D.W., Martinez, P., 1994, in *The Impact of Long-term Monitoring on Variable Star Research*, C. Sterken, M. de Groot, (eds.), Kluwer Academic Publishers, Dordrecht, 185
Kurtz, D.W., Marang, F., 1995, *Delta Scuti Star Newsletter* 8, 2
Kurtz, D.W., Kanaan, A., Martinez, P., 1993a, *MNRAS* 260, 364
Kurtz, D.W., Martinez, P., Ashley, R.P., 1993b, *MNRAS* 264, 529
Kurtz, D.W., Kreidl, T.J., O'Donoghue, D., Osip, D.J., Tripe, P., 1991, *MNRAS* 251, 152
Kwee, K.K., 1967, *Bull. Astron. Inst. Neth.* 19, 260
Kwee, K.K., Braun, L.C., 1967, *Bull. Astron. Inst. Neth. Suppl.* 2, 77
Lamers, H.J.G.L.M., 1989, in *Physics of Luminous Blue Variables*, Davidson, K., Moffat, A.F.J., Lamers, H.J.G.L.M. (eds.), Kluwer Academic Publishers, Dordrecht, 135
Landolt, A., 1986, *Ap. J.* 153, 151
Landstreet, J. D., 1992, *Astr. Ap. Rev.* 4, 35
Laney, C.D., Stobie, R.S., 1992, *Astr. Ap. Suppl.* 93, 93
Larsson-Leander, G., 1969, *Ark. f. Astron.* 5, 253
Le Contel, J.-M., Sareyan, J.-P., Valtier, J.-C., 1981, in *Workshop on Pulsating B Stars*, G. Evon, C. Sterken (eds.), Observatoire de Nice, 45
Leitherer, C., Appenzeller, I., Klare, G., Lamers, H.J.G.L.M., Stahl, O., Waters, L.B.F.M., Wolf, B., 1985, *Astr. Ap.* 153, 168
Lesh, J.R., Aizenman, M.L., 1978, *Ann. Rev. Astr. Ap.* 16, 215
Lewin, W.H.G., van den Heuvel, E.P.J. (eds.) 1983, *Accretion Driven Stellar X-ray Sources*, Cambridge University Press, Cambridge
Lines, H.C., Lines, R.D., Kirkpatrick, J.D., Hall, D.S., 1987, *Astron. J.* 93, 430
Lines, H.C., Lines, R.D., McFaul, T.G., 1988, *Astron. J.* 95, 1505
Linsky, J.L., 1980, *Ann. Rev. Astr. Ap.* 18, 439
Lipunova, N.A., 1990, in *Variable Star Research: An International Perspective*, Percy, J.R., Mattei, J.A., Sterken, C. (eds.), Cambridge University Press, Cambridge, 55
Lockwood, G.W., Wing, R.F., 1982, *MNRAS* 198, 385
Lortet, M.-C., Spite, F., 1986, *Astr. Ap. Suppl.* 64, 329
Lortet, M.-C., Borde, S., Ochsenbein, F., 1994, *Astr. Ap. Suppl.* 107, 193 (the complete paper edition appeared as *Publication Spéciale du CDS* 24, Observatoire Astronomique de Strasbourg)
Lub, J., Pel, J.W., 1977, *Astr. Ap.* 54, 137
Ludwig, K., Meyer-Hofmeister, E., Ritter, H., 1995, *Astr. Ap.* in press
Lührs, S., 1991, in *Ein geometrisch-physikalisches Umströmungsmodell für die Sternwinde in WR 79 und in verwandten Systemen*, Ph.D. Thesis, University of Münster, Germany
Lynds, C.R., 1960, *Ap. J.* 131, 122
Lyne, A.G., Graham Smith, F., 1990, in *Pulsar Astronomy*, Cambridge University Press, Cambridge
Madore, B.F. (ed.), 1985, *Cepheids; Observation and Theory*, Cambridge University Press, Cambridge
Maeder, A., 1982, *Astr. Ap.* 105, 149
Maeder, A., 1989, in *Physics of Luminous Blue Variables*, Davidson, K., Moffat, A.F.J., Lamers, H.J.G.L.M. (eds.), Kluwer Academic Publishers, Dordrecht, 15
Maeder, A., Rufener, F., 1972, *Astr. Ap.* 20, 437
Maeder, A., Conti, P.S., 1994, *Ann. Rev. Astr. Ap.*, in press
Manfroid, J., Sterken, C., 1987, *Astr. Ap. Suppl.* 71, 272

Manfroid, J., Sterken, C., 1992, *Astr. Ap.* 258, 600
Manfroid, J., Sterken, C., Bruch, A., Burger, M., de Groot, M., Duerbeck, H.W., Duemmler, R., Figer, A., Hageman, T., Hensberge, H., Jorissen, A., Madejsky, R., Mandel, H., Ott, H.-A., Reitermann, A., Schulte-Ladbeck, R.E., Stahl, O., Steenman, H., vander Linden, D., Zickgraf, F.-J., 1991a, in *First Catalogue of Stars Measured in the Long-Term Photometry of Variables Project (1982–1986)*, ESO Scientific Report No.8
Manfroid, J., Sterken, C., Bruch, A., Burger, M., de Groot, M., Duerbeck, H.W., Duemmler, R., Figer, A., Hageman, T., Hensberge, H., Jorissen, A., Madejsky, R., Mandel, H., Ott, H.-A., Reitermann, A., Schulte-Ladbeck, R.E., Stahl, O., Steenman, H., vander Linden, D., Zickgraf, F.-J., 1991b, *Astr. Ap. Suppl.* 87, 481
Manfroid, J., Sterken. C., Cunow, B., de Groot, M., Jorissen, A., Kneer, R., Krenzin, R., Kruijswijk, M., Naumann, M., Niehues, M., Schöneich, W., Sevenster, M., Vos, N., and Vogt, N., 1994a, in *Third Catalogue of Stars Measured in the Long-Term Photometry of Variables Project (1990–1992)*, ESO Scientific Report No.15
Manfroid, J., Sterken. C., Cunow, B., de Groot, M., Jorissen, A., Kneer, R., Krenzin, R., Kruijswijk, M., Naumann, M., Niehues, M., Schöneich, W., Sevenster, M., Vos, N., and Vogt, N., 1994b, *Astr. Ap. Suppl.* 109, 329
Marang, F., Carter, B.S., Roberts, G., Catchpole, R.M., Chapman, J., 1994, *MNRAS*, 267, 711
Mason, K.O., 1988, in *Multiwavelength Astrophysics*, Cordova, F. (ed.), Cambridge University Press, Cambridge
Mathias, P., Gillet, D., Kaper, L., 1991, in *Rapid Variability of OB-stars: Nature and Diagnostic Value*, ESO Workshop, Baade, D. (ed.) 193
Mathys, G., 1985, *Astr. Ap.* 151, 315
Mathys, G., 1989, *Fund. Cosmic Phys.* 13, 143
Mathys, G., 1990, *Astr. Ap.* 232, 151
Matthews, J., 1991, *Publ. Astron. Soc. Pac.* 103, 5
Mauche, C.W. (ed.) 1990, *Accretion-Powered Compact Binaries*, Cambridge University Press, Cambridge
Mauder, H., Høg, E., 1987, *Astr. Ap.* 185, 349
Mayall, M.W., 1949, *Harvard Bull.*, 919, 15
McClintock, J.E., Remillard, R.E., Margon, B., 1981, *Ap. J.* 243, 900
McHardy, I.M., Pye, J.P., Fairall, A.P., Warner, B., Cropper, M., Allen, S., 1984, *MNRAS* 210, 663
Mennickent, R.E., Vogt, N., Sterken, C., 1994, *Astr. Ap. Suppl.*, 108, 237
Meyer, F., Meyer-Hofmeister, E., 1981, *Astr. Ap.* 104, L10
Meyer, F., Meyer-Hofmeister, E., 1983, *Astr. Ap.* 121, 29
Mignard, F., Froeschlé, M., Falin, J.L., Andreasen, G.K., Høg, E., Grewing, M., Scales, D.R., 1989, ESA SP-111, Vol. III, 205
Mikolajewska, J., Friedjung, M., Kenyon, S.J., Viotti, R. (eds.) 1988, *The Symbiotic Phenomenon*, Kluwer Academic Publishers, Dordrecht
Mirzoyan, L.V., Pettersen, B.R., Tsvetkov, M.K. (eds.) 1989, *Flare Stars in Star Clusters, Associations and the Solar Vicinity*, IAU Symposium no. 137
Mochnacki, S.W., 1985, in *Interacting Binaries*, Eggleton, P.P., Pringle, J.E. (eds.), Reidel Academic Publishers, Dordrecht, 51
Moffat, A.F.J., 1994, in *The Impact of Long-Term Monitoring on Variable-Star Research*, NATO ARW, Sterken, C., de Groot, M. (eds), NATO ASI Series C, 436, Kluwer Academic Publishers, Dordrecht, 117
Moffat, A.F.J., Shara, M.M., 1986, *Astron. J.* 92, 952

Moffat, A.F.J., Drissen, L., Robert, C., 1989, in *Physics of Luminous Blue Variables*, Davidson, K., Moffat, A.F.J., Lamers, H.J.G.L.M. (eds.), Kluwer Academic Publishers, Dordrecht, 229
Moffett, T.J., 1974, *Sky Telesc.* 48, 94
Moffett, T.J., Barnes, T.G., 1980, *Ap. J. Suppl.* 44, 427
Moffett, T.J., Barnes, T.G., 1984, *Ap. J. Suppl.* 55, 389
Mohin, S., Raveendran, A.V., Mekkaden, M.V., Hall, D.S., Henry, G.W., Lines, R.D., Fried, R.E., Louth, H., Stelzer, H.J., 1985, *Ap. Sp. Sci.* 115, 353
Molnar, L. A., Kobulnicky, H.A., 1992, *Astr. Ap.* 392, 678
Morgan, W.W., 1933, *Ap. J.* 77, 330
Morris, S.L., 1985, *Ap. J.* 295, 143
Moskalik, P., Buchler, J.R., Marom, A., 1992, *Ap. J.* 385, 685
Motch, G., Ilovaisky, S.A., Chevalier, C., 1982, *Astr. Ap.* 109, L1
Müller, G., Hartwig, E., 1918, *Geschichte und Literatur des Lichtwechsels*, Erster Band, Poeschel & Trepte, Leipzig
Müller, G., Hartwig, E., 1920, *Geschichte und Literatur des Lichtwechsels*, Zweiter Band, Poeschel & Trepte, Leipzig
Müller, G., Hartwig, E., 1922, *Geschichte und Literatur des Lichtwechsels*, Dritter Band, Poeschel & Trepte, Leipzig
Munari, U., Yudin, B.F., Taranova, O.G., Massone, G., Marang, F., Roberts, G., Winkler, H., Whitelock, P.A., 1992, *Astr. Ap. Suppl.* 93, 383
Mürset, H., Nussbaumer, H., 1994, *Astr. Ap.* 282, 586
Nakagiri, M., Yamashita, Y., 1979, *Ann. Tokyo Obs.* 2nd ser., 17, 221
Nather, R.E. Robinson, E.L., 1974, *Ap. J.* 190, 637
Nemec, J.M., 1989, in *The Use of Pulsating stars in Fundamental Problems of Astronomy*, Schmidt, E.G. (ed.), Cambridge University Press, Cambridge, 215
Nemec, J.M., 1992, in *The Astronomy and Astrophysics Encyclopedia*, Maran, S.P. (ed.), Cambridge University Press, Cambridge, 789
Nemec, J.M., Lutz, T.E., 1993 in *New Perspectives on Stellar Pulsation and Pulsating Variable Stars*, Nemec, J.M., Matthews, J.M., (eds.), Cambridge University Press, Cambridge, 31
Neugebauer, G., Westphal, J.A., 1968, *Ap. J.* 152, L89
Newcomb, S., 1901, *The Stars* 116
Nolthenius, R., 1991, *Inf. Bull. Var. Stars* 3589
Olson, E.C., 1985, in *Interacting Binaries*, Eggleton, P.P., Pringle, J.E. (eds.), Reidel Academic Publishers, Dordrecht, 127
Osaki, Y., 1987, in *Instabilities in Luminous Early Type Stars*, Lamers, H.J.G.L.M., de Loore, C.W.H. (eds.), Reidel Academic Publishers, Dordrecht, 39
Osaki, Y., 1995, *Publ. Astron. Soc. Japan*, in press
Osaki, Y., Shibahashi, H., 1986, *Ap. Sp. Sci.* 118, 195
Oskanyan, V.S., Evans, D.S., Lacy, C.M., McMillan, R.S., 1977, *Ap. J.* 214, 430
Pacini, F., 1967, *Nature* 216, 567
Paczýnski, B., 1986, *Ap. J.* 304, 1
Patterson, G.J., McGraw, J.T., Coleman, L., Africano, J.L., 1981, *Ap. J.* 248, 1067
Paul, H.M., 1888, *Astron. J.* 9, 180
Payne, C.H., 1933, *Zeitschr. f. Astrophys.* 7, 1
Payne-Gaposchkin, C., 1957, *The Galactic Novae*, North-Holland Publ., Amsterdam
Payne-Gaposchkin, C., 1978, *Ann. Rev. Astr. Ap.* 16, 1
Pedersen, H., van Paradijs, J., Motch, C., Cominsky, L., Lawrence, A., Lewin, W.H., Oda, H., Ohashi, T., Matsuoka, M., 1982, *Ap. J.* 263, 340

Pel, J.W., 1976, *Astr. Ap. Suppl.* 24, 413

Penfold, J.E., Warren, P.P., Penny, A.J., 1975, *MNRAS* 171, 445

Percy, J.R., 1987, in *Physics of Be stars*, Slettebak, A., Snow, T.P. (eds.), Cambridge University Press, Cambridge, 182

Petersen, J.O., 1993, in *New Perspectives on Stellar Pulsation and Pulsating Variable Stars*, Nemec, J.M., Matthews, J.M. (eds.), Cambridge University Press, Cambridge, 371

Petersen, J.O., Andreasen, G.K., 1987, *Astr. Ap.* 176, 183

Petersen, J.O., Diethelm, R., 1986, *Astr. Ap.* 156, 337

Petersen, J.O., Hansen, L., 1984, *Astr. Ap.* 134, 319

Petschek, A.G. (ed.) 1990, *Supernovae*, Springer, Berlin

Petterson, B.R., Panov, K.P., Sandmann, W.H., Ivanova, M.S., 1986, *Astr. Ap. Suppl.* 66, 235

Pickering, E.C., 1881, *Proc. Amer. Acad. Arts & Sci.* 16, 257

Plavec, M.J., 1980, in *Close Binary Stars: Observations and Interpretation*, Plavec, M.J., Popper, D.M., Ulrich, R.K., Reidel Academic Publishers, Dordrecht, 251

Plavec, M.J., 1983, *Sky Telesc.* 65, 413

Plavec, M.J., 1985, in *Interacting Binaries*, Eggleton, P.P., Pringle, J.E. (eds.) Reidel Academic Publishers, Dordrecht, 155

Plummer, H.C., 1913, *MNRAS* 73, 661

Polyakova, T.A., 1975, *Perem. Zvezdy* 20, 75

Polyakova, T.A., 1982, *Vestnik, Ser. Mat. mekh. astron* 7, 92

Polyakova, T.A., 1984, *Astrofizika* 21, 125

Popper, D.M., Hill, G., 1991, *Astron. J.* 101, 600

Prager, R., 1934, *Geschichte und Literatur des Lichtwechsels der veränderlichen Sterne*, Zweite Ausgabe, Erster Band, Dümmlers Verl., Berlin

Prager, R., 1936, *Geschichte und Literatur des Lichtwechsels der veränderlichen Sterne*, Zweite Ausgabe, Zweiter Band, Dümmlers Verl., Berlin

Prager, R., 1941, *History and Bibliography of the Light Variations of Variable Stars*, Second Edition, Supplementary Volume, Annals Harvard College Observatory Vol. 111, Cambridge/Mass

Preston, G.W., Krzeminski, W., Smak, J., Williams, J.A., 1963, *Ap. J.* 137, 401

Reipurth, B., 1989, (ed.) *ESO Workshop on Low Mass Star Formation and Pre-Main Sequence Objects*, ESO Conference and Workshop Proceedings 33

Renson, P., Manfoid, J., 1992, *Astr. Ap.* 256, 104

Richter, G. A., Braeuer, H.-J., 1989, *Astron. Nachr.* 309, 413

Ritter, H., 1990, *A catalogue of cataclysmic binaries, low-mass X-ray binaries and related objects*, 5th ed., *Astr. Ap.* 85, 1179

Ritter, H., Kolb, U., 1995, in *X-ray Binaries*, Lewin, W.H.G., van Paradijs, J., van den Heuvel, E.P.J. (eds.), Cambridge University Press, Cambridge

Robinson, E.L., 1973, *Ap. J.* 180, 121

Roessiger, S., 1992, *Bull. Inf. CDS* 40, 39

Roessiger, S., Braeuer, H.-J., 1992, *Bibliographic Catalogue of Suspected Variable Stars (BCSVS) Update*, CDS-Archive VI/58

Roessiger, S., Braeuer, H.-J., 1993, *Bull. Inf. CDS* 42, 31

Roessiger, S., Braeuer, H.-J., 1994, *Bibliographic Catalogue of Variable Stars (BCVS), Part II (update 1992)*, CDS-Archive VI/68

Roessiger, S., Braeuer, H.-J., 1995, *Bull. Inf. CDS*, in preparation

Rosendhal, J.D., Snowden, M.S., 1971, *Ap. J.* 169, 281

Rosino, L., Taffara, S., Pinto, G., 1960, *Mem. Soc. Astron. Ital.* 31, 251

Rucinski, S.M., 1993, in *The Realm of Interacting Binary Stars*, Sahade, J., McCluskey, G.E., Kondo, Y. (eds.) 1993, Kluwer Academic Publishers, Dordrecht, 111

Rufener, F., 1988, *Catalogue of Stars measured in the Geneva Observatory Photometric system* (fourth edition), Genève

Rydgren, A.E., Vrba, F.J., 1983, *Ap. J.* 267, 191

Sahade, J., Wood, F.B., 1978, *Interacting Binary Stars*, Pergamon, Oxford

Samus', N.N., 1990, in *Variable Star Research: An International Perspective*, Percy, J.R., Mattei, J.A., Sterken, C. (eds.), Cambridge University Press, Cambridge, 52

Schiller, K., 1923, *Einführung in das Studium der veränderlichen Sterne*, J.A. Barth Verlag, Leipzig

Schneller, H., 1952, *Geschichte und Literatur des Lichtwechsels der veränderlichen Sterne*, Zweite Ausgabe, Dritter Band, Akademie-Verlag, Berlin

Schneller, H., 1957, *Geschichte und Literatur des Lichtwechsels der veränderlichen Sterne*, Zweite Ausgabe, Vierter Band, Akademie-Verlag, Berlin

Schneller, H., 1960, *Geschichte und Literatur des Lichtwechsels der veränderlichen Sterne*, Zweite Ausgabe, Fünfter Band, Erstes Heft, Akademie-Verlag, Berlin

Schneller, H., 1961, *Geschichte und Literatur des Lichtwechsels der veränderlichen Sterne*, Zweite Ausgabe, Fünfter Band, Zweites Heft, Akademie-Verlag, Berlin

Schneller, H., 1963, *Geschichte und Literatur des Lichtwechsels der veränderlichen Sterne*, Zweite Ausgabe, Fünfter Band, Drittes Heft, Akademie-Verlag, Berlin

Secchi, A., 1866, *Astron. Nachr.* 68, 63

Shara, M.M., 1989, *Publ. Astron. Soc. Pac.* 101, 5

Shibahashi, H., 1987, *Lect. Notes Phys.* 274, 112

Shibahashi, H., Takata, M., 1993, *Publ. Astron. Soc. Japan* 45, 617

Shklovski, I., 1967, *Ap. J.* 148, L1

Shobbrook, R.M., Shobbrook, R.R., 1993, *The Astronomy Thesaurus*, Anglo-Australian Observatory, Australia

Shobbrook, R.R., Stobie, R.S., 1974, *MNRAS* 169, 643 (Variable star file number IAU (27) RAS–31)

Simon, N.R., 1986, *Ap. J.* 311, 305

Simon, N.R., 1988, in *Pulsation and Mass Loss in Stars*, Stalio, R., Willson, L.A. (eds.), Kluwer Academic Publishers, Dordrecht, 27

Slettebak, A., 1979, *Space Sci. Rev.* 23, 541

Smak, J., 1993, *Acta Astron.* 43, 121

Smale, A.P., Mukai, K., 1988. *MNRAS* 231, 663

Smith, F.G., Disney, M.J., Hartley, K.F., Jones, D.H.P., King, D.J., Wellgate, G.B., Manchester, R.N., Lyne, A.G., Goss, W.M., Wallace, P.T., Peterson, B.A., Murdin, P.G., Danziger, I.J., 1978, *MNRAS* 184, 39

Smith, L.J., 1968, *MNRAS* 138, 109

Smith, L.J., 1993, in *Circumstellar Matter in the Late Stages of Stellar Evolution*, Clegg, R.E.S., Meikle, P., Stevens, I.R. (eds.), Cambridge University Press, Cambridge

Smith, M.A., 1977, *Ap. J.* 215, 574

Smith, M.A., 1980, in *Nonradial and Nonlinear Stellar Pulsation*, Hill, H.A., Dziembowski, W.A. (eds.), *Lect. Notes Phys.* Springer, Berlin 125, 105

Smith, M.A., 1994a, in *Pulsation, Rotation and Mass Loss in Early-type Stars*, Balona, L.A., Henrichs, H.F., Le Contel, J.M. (eds.), Kluwer Academic Publishers, Dordrecht, 12

Smith, M.A., 1994b, in *Pulsation, Rotation and Mass Loss in Early-type Stars*, Balona, L.A., Henrichs, H.F., Le Contel, J.M. (eds.), Kluwer Academic Publishers, Dordrecht, 241

Soderhjelm, S., 1980, *Astr. Ap.* 89, 100.
Spoon, H.W.W., de Koter, A., Sterken, C., Lamers, H.J.G.L.M., Stahl, O., 1994, *Astr. Ap. Suppl.* 106, 141
Srinivasan, G., 1989, *Astr. Ap. Rev.* 1, 209
Stagg, C., 1987, *MNRAS* 227, 213
Stahl, O., 1989, in *Physics of Luminous Blue Variables*, Davidson, K., Moffat, A.F.J., Lamers, H.J.G.L.M. (eds.), Kluwer Academic Publishers, Dordrecht, 149
Stahl, O., Wolf, B., Zickgraf, F.-J., 1987, *Astr. Ap.* 184, 193
Stebbins, J., 1923, *Ap. J.* 23, 1
Štefl, S., Baade, D., Harmanec, P., Balona, L.A., 1995, *Astr. Ap.* 294, 135
Sterken, C., 1977, *Astr. Ap.* 57, 361
Sterken, C., 1983, *Messenger* 33, 10
Sterken, C., 1988, in *Coordination of Observational Projects in Astronomy*, Jaschek, C., Sterken, C. (eds.), Cambridge University Press, Cambridge 3
Sterken, C., 1989, in *Physics of Luminous Blue Variables*, Davidson, K., Moffat, A.F.J., Lamers, H.J.G.L.M. (eds.), Kluwer Academic Publishers, Dordrecht, 59
Sterken, C., 1992, in *Vistas in Astronomy* 35, 139
Sterken, C., 1993, in *Precision Photometry*, Kilkenny, D., Lastovica, E., Menzies, J.W. (eds.), SAAO, 57
Sterken, C., 1994, in *The Impact of Long-Term Monitoring on Variable-Star Research*, NATO ARW, Sterken, C., de Groot, M. (eds), NATO ASI Series C, 436, Kluwer Academic Publishers, Dordrecht 1
Sterken, C., Jerzykiewicz, M., 1994, *Space Sci. Rev.* 62, 95
Sterken, C., Manfroid, J., 1992, *Astronomical Photometry, a Guide*, Kluwer Academic Publishers, Dordrecht
Sterken, C., Manfroid, J., 1996, *Astr. Ap.* 305, 481
Sterken, C., Vogt, N., 1995, *JAD* 1, 1
Sterken, C., Snowden, M., Africano, J., Antonelli, P., Catalano, F.A., Chahbenderian, M., Chavarria, C., Crinklaw, G., Cohen, H.L., Costa, V., de Lara, E., Delgado, A.J., Ducatel, D., Fried, R., Fu, H.-H., Garrido, R., Gilles, K., Gonzalez, S., Goodrich, B., Haag, C., Hensberge, H., Jung, J.H., Lee, S.-W., Le Contel, J.-M., Manfroid, J., Margrave, T., Naftilan, S., Peniche, R., Peña, J.H., Ratajczyk, S., Rolland, A., Sandmann, W., Sareyan, J.-P., Szuskiewicz, E., Tunca, Z., Valtier, J.-C., vander Linden, D., 1986, *Astr. Ap. Suppl.* 66, 11
Sterken, C., Young, A., Furenlid, I., 1987, *Astr. Ap.* 177, 150
Sterken, C., Gosset, E., Jüttner, A., Stahl, O., Wolf, B., Axer, M., 1991, *Astr. Ap.* 247, 383
Sterken, C., Manfroid, J., Anton, K., Barzewski, A., Bibo, E., Bruch, A., Burger, M., Duerbeck, H.W., Duemmler, R., Heck, A., Hensberge, H., Hiesgen, M., Inklaar, F., Jorissen, A., Juettner, A., Kinkel, U., Liu Zongli, Mekkaden, M.V., Ng, Y.K., Niarchos, P., Püttmann, M., Szeifert, T., Spiller, F., van Dijk, R., Vogt, N., 1993a, in *Second Catalogue of Stars Measured in the Long-Term Photometry of Variables Project (1986–1990)*, ESO Scientific Report No. 12
Sterken, C., Manfroid, J., Anton, K., Barzewski, A., Bibo, E., Bruch, A., Burger, M., Duerbeck, H.W., Duemmler, R., Heck, A., Hensberge, H., Hiesgen, M., Inklaar, F., Jorissen, A., Juettner, A., Kinkel, U., Liu Zongli, Mekkaden, M.V., Ng, Y.K., Niarchos, P., Püttmann, M., Szeifert, T., Spiller, F., van Dijk, R., Vogt, N., 1993b, *Astr. Ap. Suppl.* 102, 79
Sterken, C., Vogt, N., Mennickent, R.E., 1994, *Astr. Ap.*, 291, 473
Sterken, C., Stahl, O., Wolf, B., Szeifert, Th., Jones, A., 1995a, *Astr. Ap.* 303, 766

Sterken, C., Manfroid, J., Beele, D., de Koff, S., Eggenkamp, I.M.M.G., Göcking, K., Jorissen, A., Kaufer, A., Kuss, C., Schoenmakers, A.P., Stil, J.M., van Loon, J., Vink, J., Vrielmann, S., Wälde, E., 1995b, in *Fourth Catalogue of Stars Measured in the Long-Term Photometry of Variables Project*. ESO Scientific Report No. 14

Sterken, C., Manfroid, J., Beele, D., de Koff, S., Eggenkamp, I.M.M.G., Göcking, K., Jorissen, A., Kaufer, A., Kuss, C., Schoenmakers, A.P., Stil, J.M., van Loon, J., Vink, J., Vrielmann, S., Wälde, E., 1995c, *Astr. Ap. Suppl.* 113, 31

Sterken, C., de Groot, M.W.H., van Genderen, A.M., 1996, *Astr. Ap.* in press

Sterken, C., Vogt, N., Mennickent, R.E., 1996, *Astr. Ap.*, in press

Stibbs, D.W.N., 1950, *MNRAS* 110, 395

Stobie, R.S., 1973, *Observatory* 93, 111

Strassmeier, K.G., Hall, D.S., 1988, *Ap. J. Suppl.* 67, 453

Strassmeier, K.G., Hall, D.S., Fekel, F.C., Scheck, M., 1993, *Astr. Ap. Suppl.* 100, 173

Suntzeff, N.B., Hamuy M., Martin G., Gómez A., Gonzalez, R., 1988, *Astron. J.* 96, 1864

Szkody, P., Mattei, J.A., 1984, *Publ. Astron. Soc. Pac.* 96, 988

Taylor, J.H., Stinebring, D.R., 1986, *Ann. Rev. Astr. Ap.* 24, 285

Teays, T.J., 1993, in *New Perspectives on Stellar Pulsation and Pulsating Variable Stars*, Nemec, J.M., Matthews, J.M., (eds.), Cambridge University Press, Cambridge 410

Thomas, H.C., 1977, *Ann. Rev. Astr. Ap.* 15, 127

Udalski, A., 1988, *Acta Astron.* 38, 315

Unno, W., Osaki, Y., Ando, H., Saio, H., Shibahashi, H., 1988, *Nonradial Oscillations of Stars*, University of Tokyo Press

Vaidya, J., Agrawal, P.C., Apparao, K.M., Manchanda, R.K., Vivekananda Rao, P., Sarma, M.B.K., 1988, *Astr. Ap. Suppl.* 75, 43

van den Bergh, S., 1994, *Ap. J. Suppl.* 92, 219

van den Bergh, S., Tammann, G.A., 1991, *Ann. Rev. Astr. Ap.* 29, 363

van den Heuvel, E.P.J., Rappaport, S.A., 1987 in *Physics of Be Stars*, IASU Coll. 92, Slettebak, A., Snow, P. (eds.), Kluwer Academic Publishers, Dordrecht, 291

van der Hucht, K.A., 1992, *Astr. Ap. Rev.* 4, 123

van der Hucht, K.A., Conti, P.S., Lundström, I., Stenholm, B, 1981, *Space Sci. Rev.* 28, 227

van der Hucht, K.A., Williams, P.M., Spoelstra, T.A.Th., Swaanenvelt J.P., 1994, *The Impact of Long-Term Monitoring on Variable-Star Research*, NATO ARW, Sterken, C., de Groot, M. (eds), NATO ASI Series C, 436, Kluwer Academic Publishers, Dordrecht, 75

van Genderen, A.M., 1979, *Astr. Ap. Suppl.* 38, 151

van Genderen, A.M., 1995, private communication

van Genderen, A.M., Thé, P.S., 1984, *Space Sci. Rev.* 39, 317

van Genderen, A.M., van Amerongen, S., van der Bij, M.D.P., Damen, E., van Driel, W., Greve, A., van Heerde, G.M., Latour, H.J., Ng, Y.K., Oppe, J., Wiertz, M.J.J., 1988, *Astr. Ap. Suppl.* 74, 467

van Genderen, A.M., Steemers, W.J.G., Feldbrugge, P.T.M. et al., 1985, *Astr. Ap.* 153, 163

van Genderen, A.M., Thé, P.S., Augusteijn, Th. et al., 1988, *Astr. Ap. Suppl.* 74, 453

van Genderen, A.M., Thé, P.S., Heemskerk, M., et al., 1990a, *Astr. Ap.* 82, 189

van Genderen, A.M., van der Hucht, K.A., Larsen, I., 1990b, *Astr. Ap.* 229, 123

van Genderen, A.M., de Groot, M.J.H., Thé, P.S., 1994, *Astr. Ap.* 283, 89

van Genderen, A.M., van den Bosch, F.C., Dessing, F., Fehmers, G.C., van Grunsven, J., van der Heiden, R., Janssens, A.M., Kalter, R., van der Meer, R.L.J., van Ojik, R., Smit, J.M., Zijderveld, M.J., 1992, *Astr. Ap.* 264, 88
van Genderen, A.M., Sterken, C., de Groot, M.J.H., Stahl, O., Andersen, J., Andersen, M.I., Caldwell, J.A.R., Casey, B., Clement, R., Corradi, W.J.B., Cuypers, J., Debehogne, H., Garcia de Maria, J.M., Jönch-Sörensen, H., Vaz, L.P. R., Štefl, S., Suso Lopez, J., Beele, D., Eggenkamp, I.M.M.G., Göcking, K.-D., Jorissen, A., de Koff, S., Kuss, C., Schoenmakers, A., Vink, J., Wälde, E., 1995, *Astr. Ap.* 304, 415
van Gent, R.H., Lamers, H.J.G.L.M., 1986, *Astr. Ap.* 158, 335
van Paradijs, J., 1983, in *Accretion Driven Stellar X-ray Sources*, Lewin, W.H.G., van den Heuvel, E.P.J. (eds.), Cambridge University Press, Cambridge, 189
van 't Veer, F., 1991, *Astr. Ap.* 250, 84
vander Linden, D., Sterken C., 1985, *Astr. Ap.* 150, 76
vander Linden, D., Sterken, C., 1987, *Astr. Ap.* 186, 129
Ventura, J., Pines, D. (eds.), 1991, *Neutron Stars, Theory and Observations*, Kluwer Academic Publishers, Dordrecht
Vetesnik, M., 1984, *Bull. Astron. Inst. Czech.* 35, 74
Viotti, R., 1990, in *Variable Star Research: An International Perspective*, Percy, J.R., Mattei, J.A., Sterken, C. (eds.), Cambridge University Press, Cambridge, 194
Vogt, N., 1974, *Astr. Ap.* 36, 369
Vogt, N., 1982, *Ap. J.* 252, 653
Vogt, N., 1995, in *Cataclysmic Variables: Inter Class Relations*, della Valle, M. (ed.), Proceedings of the Padova-Abano Conference, in press
Vogt, N., Breysacher, J., 1980, *Ap. J.* 235, 945
Vogt, S.S., Penrod, G.D., 1983, *Ap. J.* 275, 661
Vreux, J.M., 1987, in *Instabilities in Luminous Early Type Stars*, Lamers, H.J.G.L.M., de Loore, C.W.H. (eds.), Reidel Academic Publishers, Dordrecht, 81
Waelkens, C., 1991a, *Astr. Ap.* 244, 107
Waelkens, C., 1991b, *Astr. Ap.* 246, 453
Waelkens, C., Rufener, F., 1985, *Astr. Ap.* 152, 6
Wallerstein, G., Cox, A.N., 1984, *Publ. Astron. Soc. Pac.* 96, 677
Walraven, Th., 1949, *Bull. Astron. Inst. Neth.* 11, 17
Walraven, Th., 1955, *Bull. Astron. Inst. Neth.* 12, 223
Walraven, Th., Walraven, J., Balona, L.A., 1992, *MNRAS* 254, 59
Walter K., Duerbeck, H.W., 1988, *Astr. Ap.* 189, 89
Warner, B., 1972, *MNRAS* 159, 315
Warner, B., 1995, *Cataclysmic Variable Stars, an Introduction to Observational Properties and Physical Structure*, Cambridge University Press, Cambridge
Warner, B., Nather, R.E., 1972a, *MNRAS* 156, 1
Warner, B., Nather, R.E., 1972b, *MNRAS* 156, 297
Warner, B., O'Donoghue, D., 1988, *MNRAS* 233, 705
Warner, B., Robinson, E.L., 1972, *MNRAS* 159, 101
Warner, B., van Citters, G.W., 1974, *Observatory* 94, 116
Watson, R.D., 1988, *Ap. Sp. Sci.* 140, 255
Whelan, J.A.J., Ward, M.J., Allen, D.A., Danziger, I.J., Fosbury, R.A.E., Murdin, P.G., Penston, M.V., Peterson, B.A., Wampler, E.J., Webster, E.L., 1977, *MNRAS* 180, 657
Whitehurst, R., 1988, *MNRAS* 232, 35
Whitelock, P. A., 1987, *Publ. Astron. Soc. Pac.*, 99, 573

Whitelock, P.A., Feast, M.W., Koen, C., Roberts, G., Carter, B.S., 1994, *MNRAS* 270, 364
Whitelock, P. A., Menzies, J. W., Feast, M., Marang, F., Carter, B., Roberts, G., Catchpole, R., Chapman, J., 1994, *MNRAS* 267, 711
Willson, L.A., 1986, *JAAVSO* 15, 228
Wilson, R.E., 1989, *Space Sci. Rev.* 50, 191
Wilson, R.E., Devinney, E.,F., 1971, *Ap. J.* 166, 605
Wilson, R.E., Fox, R.K., 1981, *Astron. J.* 86, 1259
Wing, R.F., 1986, *JAAVSO* 15, 212
Winget, D.E., 1988, in *Multimode Stellar Pulsations*, Kovács, G., Szabados, L., Szeidl, B. (eds.), Konkoly Observatory/Kultura, Budapest, 181
Winget, D. E., Nather, R.E., Clemens, J.C., Provencal, J., Kleinman, S.J., Bradley, P.A., Wood, M.A., Claver, C.F., Robinson, E.L., Grauer, A.D., Hine, B.P., Fontaine, G., Achilleos, N., Marar, T.M.K., Seetha, S., Ashoka, B.N., O'Donoghue, D., Warner, B., Kurtz, D.W., Martinez, P., Vauclair, G., Chevreton, M., Kanaan, A., Kepler, S.O., Augusteijn, T., van Paradijs, J., Hansen, C.J., Liebert, J., 1990, *Ap. J.* 357, 630
Winiarski, M., Zola, S., 1987, *Astr. Ap.* 37, 375
Wisniewski, W.Z., 1973, *MNRAS* 161, 331
Wolf, B., 1986, in *Luminous Stars and Associations in Galaxies*, de Loore, C.W.H., Willis, A.J., Laskarides, P. (eds.) Reidel Academic Publishers, Dordrecht, 151
Wolf, B., 1989, in *Physics of Luminous Blue Variables*, Davidson, K., Moffat, A.F.J., Lamers, H.J.G.L.M. (eds.), Kluwer Academic Publishers, Dordrecht, 91
Wolf, B., 1992, in *Non-isotropic and Variable Outflows from Stars*, Drissen, L., Leitherer, C., Notta, A. (eds.), *Astron. Soc. Pac. Conf. Ser.*, 22, 327
Wolf, B., 1994, in *The Impact of Long-Term Monitoring on Variable-Star Research*, NATO ARW, Sterken, C., de Groot, M. (eds), NATO ASI Series C, 436, Kluwer Academic Publishers, Dordrecht, 291
Wolf, B., Appenzeller, I., Stahl, O., 1981, *Astr. Ap.* 103, 94
Wolf, C.J.E., Rayet, G., 1867, *Comptes Rendus Hebdomadaires des Séances de l'Académie des Sciences* 65, 292
Wolff, S.C., 1983, *The A-type Stars: Problems and Perspective*, NASA SP-463 Monograph series on nonthermal phenomena in stellar atmospheres, Washington/Paris
Wood, F.B., Oliver, J.P., Florkowski, D.R., Koch, R.H., 1980, *A Finding List for Observers of Interacting Binary Stars* (5th Ed.), Publications of the University of Pennsylvania Astronomical Series Vol. 12, Philadelphia, University of Pennsylvania Press
Wood, J., Horne, K., Berriman, G., Wade, R., O'Donoghue, D., Warner, B., 1986, *MNRAS* 219, 629
Woosley, S.E. (ed.) 1991, *Supernovae*, 10th Santa Cruz Workshop, Springer, New York
Woosley, S.E., Weaver, T.A., 1986, *Ann. Rev. Astr. Ap.* 24, 205
Wright, E.L., Gottlieb, E.W., Liller, W., 1975, *Ap. J.* 200, 171
Young, A., 1992, in *Automatic Telescopes for Photometry and Imaging*, Adelman, S.J., Dukes, R.J., Adelman, C.J. (eds.), ASP Conference Series, 28, 73
Zanella, R., Wolf, B., Stahl, O., 1984, *Astr. Ap.* 137, 79
Zeilik, M., Hall, D.S., Feldman, P.A., Walter, F.M., 1979, *Sky Telesc.* 57, 132
Zhang, E.H., Robinson, E.L., 1987, *Ap. J.* 321, 813
Zinner, E. (von), 1926, *Veröff. Remeis-Sternwarte Bamberg* Band II.
Zwicky, F., 1974, in *Supernovae and Supernova Remnants*, Cosmovici, C.B. (ed.) Reidel Academic Publishers, Dordrecht, 1

Addresses of interest

IAU Archives of unpublished observations of variable stars
Electronic storage and retrieval of data:

- Dr E. Schmidt, University of Nebraska, Lincoln, Nebraska, USA
- Dr P. Dubois, Centre de Données Stellaires, Observatoire de Strasbourg, 11, rue de l'Université, F-670000 Strasbourg, France
- Mr P.D. Hingley, Librarian, Royal Astronomical Society, Burlington House, London, W1V 0NL, Great Britain
- Dr Yu.S. Romanov, Odessa Astronomical Observatory, Shevchenko Park, Odessa 270014, Ukraine

Central Bureau for Astronomical Telegrams (IAU Commission 6)
Smithsonian Astrophysical Observatory, Cambridge, Mass. 02138, USA

Information Bulletin on Variable Stars (IBVS)
Editors: Dr Laszlo Szabados and Dr Katalin Oláh, Konkoly Observatory, 1525 Budapest, P.O. Box 67, Hungary

American Association of Variable Stars Observers (AAVSO)
Dr J.A. Mattei, 25 Birch Street, Cambridge, Mass. 02138-1205, USA

Association Française des Observateurs d'Etoiles Variables (AFOEV)
Mr E. Schweitzer, Observatoire de Strasbourg, 11, rue de l'Université, F-67000 Strasbourg, France

Groupe Européen d'Observations Stellaires (GEOS)
3 Promenade Vénézia, F-78000 Versailles, France

Bundesdeutsche Arbeitsgemeinschaft für Veränderliche Sterne (BAV)
Munsterdamm 90, 12169 Berlin, Germany

Journal of Astronomical Data, Mr G. Kiers, TWIN Press, Prickwaert 122, 3363 SE Sliedrecht, The Netherlands
Editors: Dr C. Sterken and Dr M. de Groot

BAA – Variable Star Section
Office: Burlington House, London, W1V ONL, Great Britain

RASNZ – Variable Star Section
Dr F.M. Bateson, P.O. Box 3093, Greerton, Tauranga, New Zealand

International Amateur Professional Photoelectric Photometry (IAPPP)
Fairborn Observatory, 1247 Folk Road, Fairborn OH 45324, USA

Appendix: Tables

Table A1 List of constellation abbreviations, names and genitive forms

Ab.	Name	Genitive	Ab.	Name	Genitive
And	Andromeda	Andromedae	Lac	Lacerta	Lacertae
Ant	Antlia	Antliae	Leo	Leo	Leonis
Aps	Apus	Apodis	LMi	Leo Minor	Leonis Minoris
Aqr	Aquarius	Aquarii	Lep	Lepus	Leporis
Aql	Aquila	Aquilae	Lib	Libra	Librae
Ara	Ara	Arae	Lup	Lupus	Lupi
Ari	Aries	Arietis	Lyn	Lynx	Lyncis
Aur	Auriga	Aurigae	Lyr	Lyra	Lyrae
Boo	Boötes	Boötis	Men	Mensa	Mensae
Cae	Caelum	Caeli	Mic	Microscopium	Microscopii
Cam	Camelopardus	Camelopardalis	Mon	Monoceros	Monocerotis
Cnc	Cancer	Cancri	Mus	Musca	Muscae
CVn	Canes Venatici	Canum Venaticorum	Nor	Norma	Normae
CMa	Canis Major	Canis Majoris	Oct	Octans	Octantis
CMi	Canis Minor	Canis Minoris	Oph	Ophiuchus	Ophiuchi
Cap	Capricornus	Capricorni	Ori	Orion	Orionis
Car	Carina	Carinae	Pav	Pavo	Pavonis
Cas	Cassiopeia	Cassiopeiae	Peg	Pegasus	Pegasi
Cen	Centaurus	Centauri	Per	Perseus	Persei
Cep	Cepheus	Cephei	Phe	Phoenix	Phoenicis
Cet	Cetus	Ceti	Pic	Pictor	Pictoris
Cha	Chamaeleon	Chamaeleontis	Psc	Pisces	Piscium
Cir	Circinus	Circini	PsA	Piscis Austrinus	Piscis Austrini
Col	Columba	Columbae	Pup	Puppis	Puppis
Com	Coma Berenices	Comae Berenices	Pyx	Pyxis	Pyxidis
CrA	Corona Australis	Coronae Australis	Ret	Reticulum	Reticuli
CrB	Corona Borealis	Coronae Borealis	Sge	Sagitta	Sagittae
Crv	Corvus	Corvi	Sgr	Sagittarius	Sagittarii
Crt	Crater	Crateris	Sco	Scorpius	Scorpii
Cru	Crux	Crucis	Scl	Sculptor	Sculptoris
Cyg	Cygnus	Cygni	Sct	Scutum	Scuti
Del	Delphinus	Delphini	Ser	Serpens[1]	Serpentis
Dor	Dorado	Doradus	Sex	Sextans	Sextantis
Dra	Draco	Draconis	Tau	Taurus	Tauri
Equ	Equuleus	Equulei	Tel	Telescopium	Telescopii
Eri	Eridanus	Eridani	Tri	Triangulum	Trianguli
For	Fornax	Fornacis	TrA	Triangulum Australe	Trianguli Australe
Gem	Gemini	Geminorum	Tuc	Tucana	Tucanae
Gru	Grus	Gruis	UMa	Ursa Major	Ursae Majoris
Her	Hercules	Herculis	UMi	Ursa Minor	Ursae Minoris
Hor	Horologium	Horologii	Vel	Vela	Velorum
Hya	Hydra	Hydrae	Vir	Virgo	Virginis
Hyi	Hydrus	Hydri	Vol	Volans	Volantis
Ind	Indus	Indi	Vul	Vulpecula	Vulpeculae

1 Note that the constellation Serpens consists of two distinct parts–separated by Ophiuchus–viz. Serpens Caput (head of the snake) and Serpens Cauda (tail of the snake). Both parts, however, share the same numbering system for member stars.

Table A2 Julian Dates (*minus* 2 400 000) on January 1 of each year, at 12 h UT

	0	1	2	3	4	5	6	7	8	9
1860	00411	00777	01142	01507	01872	02238	02603	02968	03333	03699
1870	04064	04429	04794	05160	05525	05890	06255	06621	06986	07351
1880	07716	08082	08447	08812	09177	09543	09908	10273	10638	11004
1890	11369	11734	12099	12465	12830	13195	13560	13926	14291	14656
1900	15021	15386	15751	16116	16481	16847	17212	17577	17942	18308
1910	18673	19038	19403	19769	20134	20499	20864	21230	21595	21960
1920	22325	22691	23056	23421	23786	24152	24517	24882	25247	25613
1930	25978	26343	26708	27074	27439	27804	28169	28535	28900	29265
1940	29630	29996	30361	30726	31091	31457	31822	32187	32552	32918
1950	33283	33648	34013	34379	34744	35109	35474	35840	36205	36570
1960	36935	37301	37666	38031	38396	38762	39127	39492	39857	40223
1970	40588	40953	41318	41684	42049	42414	42779	43145	43510	43875
1980	44240	44606	44971	45336	45701	46067	46432	46797	47162	47528
1990	47893	48258	48623	48989	49354	49719	50084	50450	50815	51180
2000	51545	51911	52276	52641	53006	53372	53737	54102	54467	54833

Table A3 List of variable-star newsletters

Type of variable	Editor	Email address
B/Be stars	Geraldine Peters	gies@chara.gsu.edu http://chara.gsu.edu/BeNews.html
Hot stars	Philippe Eenens	eenens@tonali.inaoep.mx
Mpec giants	Sandra Yorka	yorka@cc.denison.edu
δ Scuti stars	Michel Breger	breger@astro.ast.univie.ac.at http://venus.ast.univie.ac.at
Whole Earth Telescope	Darragh O'Donaghue	wet@uctvms.uct.ac.za
Peculiar stars	Pierre North	north@scsun.unige.ch
Standard stars	Fr. Chris Corbally	corbally@as.arizona.edu
AGB stars	C. Bertout	http://gag/observ-gr.fr/liens/agbnews.html
Star formation	Bo Reipurth	fmurtagh@eso.org
Hot star winds	Steve Cranmer	http://www.bartol.udel.edu/cranmer/hotstar-home.html

Illustration credits

1.1 1989, ESA SP-111, Vol. III, 205
1.2 1990, in *Variable Star Research: An International Perspective*, Percy, J.R., Mattei, J.A., Sterken, C. (eds.), Cambridge University Press, Cambridge, 52
1.3 1990, in *Variable Star Research: An International Perspective*, Percy, J.R., Mattei, J.A., Sterken, C. (eds.), Cambridge University Press, Cambridge, 55
1.4 1992, *Astronomical Photometry, a Guide*, Kluwer Academic Publishers, Dordrecht
1.5 1977, *Astr. Ap.* 54, 137
1.6 1994, in *Applications de la photométrie de Geneève aux étoiles B et à l'extinction interstellaire*, Thèse 2692, Université de Genève
1.7 1992, in *Vistas in Astronomy* 35, 139
1.8 1987, *Astr. Ap. Suppl.* 71, 272
1.9 LTPV
1.10 1992, *Messenger* 70, 35
1.11 1992, *Messenger* 70, 35

2.1 LTPV
2.2 LTPV & 1994, *Astr. Ap. Suppl.* 106, 141
2.3 LTPV & 1994, *Astr. Ap. Suppl.* 106, 141
2.4 1985, *Astr. Ap.* 153, 163
2.5 LTPV & 1994, *Astr. Ap. Suppl.* 106, 141
2.6 1987, *MNRAS* 227, 213
2.7 1988, *Ir. Astron. J.* 18, 163
2.8 M. de Groot
2.9 1995, *Astr. Ap.* 304, 415
2.10 LTPV & 1994, *Astr. Ap. Suppl.* 106, 141
2.11 1992, *Astr. Ap.* 264, 88
2.12 1994, in *The Impact of Long-Term Monitoring on Variable-Star Research*, NATO ARW, Sterken, C., de Groot, M. (eds), NATO ASI Series C, 436, Kluwer Academic Publishers, Dordrecht, 117
2.13 1989, *MNRAS* 240, 103
2.14 1990b, *Astr. Ap.* 229, 123
2.15 1995b, *Astr. Ap.* 303, 766
2.16 1992, in *Evolutionary Processes in Interacting Binary Stars*, Kondo, Y., Sistero, R., Polidan, R.S. (eds.), Kluwer Academic Publishers, Dordrecht, 123
2.17 A. Jones
2.18 1980, *Astr. Ap.* 90, 207
2.19 1977, *Ap. J.* 217, 693
2.20 1979, *Astr. Ap.* 79, L18
2.21 1977, *Ap. J.* 217, 693
2.22 1977, *Ap. J.* 217, 693
2.23 1982, *Astron. J.* 87, 1710
2.24 1989, *Astr. Ap.* 211, 99
2.25 1989, *Astr. Ap.* 211, 115
2.26 1983, *Ap. J.* 267, 191
2.27 1980, *Astron. J.* 85, 555
2.28 1980, *Astron. J.* 85, 555
2.29 1991, *Astr. Ap. Suppl.* 89, 319
2.30 1973, *Astr. Ap. Suppl.* 10, 47
2.31 1973, *Ap. J.* 185, 239
2.32 1973, *Astr. Ap. Suppl.* 10, 47
2.33 1973, *Astr. Ap. Suppl.* 10, 47
2.34 1973, *Astr. Ap. Suppl.* 10, 47
2.35 1974, *Sky Telesc.* 48, 94
2.36 1986, *Astr. Ap. Suppl.* 66, 235
2.37 AAVSO
2.38 1972, *MNRAS* 158, 305

3.1 1988, *Astr. Ap. Suppl.* 74, 453
3.2 1992, *Astr. Ap.* 264, 88
3.3 1979, *Astr. Ap. Suppl.* 38, 151
3.4 1996, *Astr. Ap. Suppl.*, submitted
3.5 1986, *Astr. Ap. Suppl.* 66, 11
3.6 1986, *MNRAS* 219, 449
3.7 1987, *Astr. Ap.* 177, 150
3.8 1984, *MNRAS* 211, 297
3.9 1985, *Astr. Ap.* 150, 76
3.10 1979, in *Changing Trends in Variable Star Research*, (IAU Coll 46), Bateson F.M., Smak J., Urch I.H. (eds.), University of Waikato, 474
3.11 1994, in *The Impact of Long-Term Monitoring on Variable-Star Research*, NATO ARW, Sterken, C., de Groot, M. (eds), NATO ASI Series C, 436, Kluwer Academic Publishers, Dordrecht, 55
3.12 1991, *Astr. Ap.* 244, 107

3.13 1995, *Astr. Ap.* 294, 135
3.14 1994, *Astr. Ap.*, 291, 473
3.15 1996, *Astr. Ap.*, in press
3.16 1996, *Astr. Ap.* 305, 481
3.17 1985, *Astr. Ap.* 152, 6
3.18 1996, *Astr. Ap.* 308, 151
3.19 1977, *Publ. Astron. Soc. Pac.* 89, 339
3.20 1974, *MNRAS* 169, 643
3.21 1949, *Bull. Astron. Inst. Neth.* 11, 17
3.22 1987, *Ap. J.* 314, 605
3.23 1974, *Publ. Astron. Soc. Pac.* 86, 912
3.24 1981, *Ap. J. Suppl.* 47, 315
3.25 1985, *SAAO Circ.* 9, 5
3.26 1985, *SAAO Circ.* 9, 5 and 1979, *Ap. J.* 232, 197
3.27 1984, *Ap. J. Suppl.* 55, 389 and 1984, *Astr. Ap.* 134, 319
3.28 1967, *Bull. Astron. Inst. Neth. Suppl.* 2, 77
3.29 1961, *Ap. J. Suppl.* 6, 319
3.30 1967, *Bull. Astron. Inst. Neth. Suppl.* 2, 77
3.31 1963, *Ap. J.* 137, 401
3.32 AAVSO
3.33 AAVSO
3.34 1963, *Ap. J.* 137, 401
3.35 1977, *Ap. J. Suppl.* 34, 233
3.36 1977, *Ap. J. Suppl.* 34, 233
3.37 AAVSO
3.38 1952, *Harvard Annals* 118, 171
3.39 1975, *Ap. J. Suppl.* 29, 77
3.40 AAVSO
3.41 1982, *MNRAS* 198, 385
3.42 1984, *Astrofizika* 21, 125
3.43 AAVSO
3.44 AAVSO
3.45 1975, *Ap. J. Suppl.* 29, 77
1977, *Astr. Ap. Suppl.* 29, 15
1979, *Ann. Tokyo Obs.* 2nd ser., 17, 221
3.46 AAVSO
3.47 1975, *Ap. J. Suppl.* 29, 77
1977, *Astr. Ap. Suppl.* 29, 15
1973, *Ap. J. Suppl.* 25, 369
3.48 AAVSO
3.49 1972, *MNRAS* 156, 1

4.1 1993, *Astr. Ap. Suppl.* 97, 501
4.2 1992, *Astr. Ap.* 256, 104
4.3 1995, *Astr. Ap. Suppl.* 109, 425
4.4 1991, *MNRAS* 249, 468
1993, *MNRAS* 260, 364
1991, *MNRAS* 251, 152
4.5 1991, *MNRAS* 251, 152
4.6 1960, *Ap. J.* 131, 122
4.7 1980, *Ap. Sp. Sci.* 68, 358
4.8 1990, *IAPPP Comm.* 42, 44
4.9 1977, *Ap. J.* 214, 430
4.10 1959, *MNRAS* 119, 526
4.11 1981, *Publ. Astron. Soc. Pac.* 93, 504
4.12 1984, *Ap. J.* 285, 202
4.13 1990, *IAPPP Comm.* 42, 44
4.14 1978, *MNRAS* 184, 39

5.1 1973, *Astr. Ap.* 28, 295
5.2 1983, *Publ. Astron. Soc. Pac.* 95, 72
5.3 AAVSO
5.4 1988, *Astron. J.* 95, 63
5.5 1981, *Publ. Astron. Soc. Pac.* 93, 36
5.6 AAVSO
5.7 H. Duerbeck
5.8 AAVSO
5.9 AAVSO
5.10 H. Duerbeck
5.11 AAVSO
5.12 1967, *Publ. Astron. Soc. Pac.* 70, 600
1958, *MNASSA* 17, 132
1960, *Mem. Soc. Astron. Ital.* 31, 251
5.13 1972, *MNRAS* 159, 315
5.14 1972, *MNRAS* 159, 101
5.15 1985, *MNRAS* 212, 709
5.16 1984, *MNRAS* 210, 663
5.17 1988, *Astr. Ap. Suppl.* 75, 43
5.18 1974, *Ap. J.* 190, 637
5.19 1974, *Observatory* 94, 116
5.20 AAVSO
5.21 1988, *Acta Astron.* 38, 315
5.22 1995, *JAD* 1, 1
5.23 AAVSO
5.24 1987, *Ap. J.* 321, 813
5.25 AAVSO
5.26 AAVSO
5.27 1980, *Ap. J.* 235, 945
5.28 1973, *Ap. J.* 180, 121
5.29 AAVSO
5.30 1984, *Publ. Astron. Soc. Pac.* 96, 988
5.31 1977, *JBAA* 87, 395
5.32 1977, *New Zealand Journal of Science* 20, 73
5.33 1974, *Astr. Ap.* 36, 369
5.34 1974, *Astr. Ap.* 36, 369
5.35 1986, *MNRAS* 219, 629
5.36 1988, *MNRAS* 233, 705
5.37 1988, *MNRAS* 233, 705
5.38 AAVSO
5.39 AAVSO
5.40 1972, *MNRAS* 156, 297
5.41 AAVSO
5.42 1981, *Ap. J.* 248, 1067
5.43 1970, *Astrofizika* 6, 49
5.44 AAVSO
5.45 AAVSO
5.46 1991, Bateson & Kenyon, private comm.
5.47 1970, *Astrofizika* 6, 49
5.48 AAVSO

6.1 1970, *Publ. Astron. Soc. Pac.* 82, 1077
6.2 1974, *Publ. Astron. Soc. Pac.* 86, 912
6.3 1969, *Publ. Astron. Soc. Pac.* 81, 754
6.4 1979, *Acta Astron.* 29, 259
6.5 1984, *Astron. J.* 87, 559
6.6 1969, *Ark. f. Astron.* 5, 253
6.7 1969, *Astron. J.* 74, 1191
6.8 1990, *IAPPP Comm.* 42, 44
6.9 1990, *IAPPP Comm.* 42, 44
6.10 1987, *Astr. Ap.* 37, 375
6.11 1990, *IAPPP Comm.* 42, 44

6.12 1990, *IAPPP Comm.* 42, 44
6.13 1990, *IAPPP Comm.* 42, 44
6.14 1981, *Ap. J.* 251, 585
1985, *Ap. Sp. Sci.* 115, 353
6.15 1994, *Astr. Ap.* 289, 827
6.16 1988, *Astr. Ap.* 189, 89
6.17 1978, *Astr. Ap. Suppl.* 32, 361

7.1 1983, in *Accretion Driven Stellar X-ray Sources*, Lewin, W.H.G., van den Heuvel, E.P.J. (eds.), Cambridge University Press, Cambridge, 189
7.2 1983, in *Accretion Driven Stellar X-ray Sources*, Lewin, W.H.G., van den Heuvel, E.P.J. (eds.), Cambridge University Press, Cambridge, 189
7.3 1975, *MNRAS* 171, 445
7.4 1981, *Space Sci. Rev.* 30, 405
7.5 1975, *Ap. J.* 200, 171
7.6 1982, *Ap. J.* 263, 340
7.7 1982, *Ap. J.* 263, 340
7.8 1988, *MNRAS* 231, 663
7.9 1977, *MNRAS* 180, 657
7.10 1982, *Astr. Ap.* 109, L1
7.11 1981, *Ap. J.* 243, 900

Object index

Subject index